MANIFOLDS - CURRENT RESEARCH AREAS

Edited by **Paul Bracken**

Manifolds - Current Research Areas
http://dx.doi.org/10.5772/63138
Edited by Paul Bracken

Contributors

Chao Wang, Xubo Song, Yuanhao Guo, Paul Bracken, Yong Seung Cho, Xianglei Xing, Kejun Wang, Weixing Feng, Mehmet Atceken, Umit Yildirim, Qinlong Wang, Bo Sang, Wentao Huang

Published by InTech

Janeza Trdine 9, 51000 Rijeka, Croatia

Publishing Process Manager Iva Lipovic

Technical Editor SPi Global

Cover InTech Design team

First published January, 2017

Printed in Croatia

Additional hard copies can be obtained from orders@intechopen.com

Manifolds - Current Research Areas, Edited by Paul Bracken

p. cm.

Print ISBN 978-953-51-2871-7
Online ISBN 978-953-51-2872-4

For EU product safety concerns:
IN TECH d.o.o., Prolaz Marije Krucifikse Kozulić 3, 51000 Rijeka, Croatia,
info@intechopen.com or visit our website at intechopen.com.

Meet the editor

The book editor Dr. Paul Bracken is a professor in the Department of Mathematics at the University of Texas located in Edinburg, TX. He has a BSc degree from the University of Toronto and PhD degree from the University of Waterloo in Canada. His research interests include applications of differential geometry and partial differential equations to problems in mathematical physics such as gravity as well as problems in quantum theory and quantum field theory. He has published well over 100 papers in journals and books and given talks at various meetings and conferences and different levels. This is the third volume he has contributed to and published with InTech publisher.

Contents

Preface

This volume is a collection of papers on the subject of manifolds. Manifolds are an important and crucial structure in modern mathematics. They have been intensively investigated over the last 60 years and provide a foundation over which much of modern differential geometry has developed. The contributions represented here investigate manifolds of particular types, such as symplectic manifolds and submanifolds. They discuss what can be learned about manifolds by defining and studying various structures on them. In particular, there is a paper covering operator actions on manifolds and their spectral properties. There is a paper on Bonnet surfaces which emphasizes the important role played by differential equations in the study of manifolds and surfaces in particular. A paper on bifurcations and manifolds and two papers on the application of manifolds to some areas of applied mathematics are presented as well. Finally, there is a paper on symplectic affine actions on manifolds.

This book has been put together by an international group of invited authors, and it is a pleasure to thank them for their hard work and significant contributions to this volume. I gratefully acknowledge with great thanks the assistance and help provided by Ms. Iva Lipovic who was the publishing manager throughout this process as well as the InTech publishing group for the opportunity to edit this volume on the subject of manifolds.

Paul Bracken
Department of Mathematics,
University of Texas,
Edinburg, TX, USA

Section 1

Hopf Bifurcations

Chapter 1

Mutiple Hopf Bifurcation on Center Manifold

Qinlong Wang, Bo Sang and Wentao Huang

Additional information is available at the end of the chapter

http://dx.doi.org/10.5772/65674

Abstract

In this chapter, by researching the algorithm of the formal series, and deducing the recursion formula of computing the nondegenerate and degenerate singular point quantities on center manifold, we investigate the Hopf bifurcation of high-dimensional nonlinear dynamic systems. And more as applications, the singular point quantities for two classes of typical three- or four-dimensional polynomial systems are obtained, the corresponding multiple limit cycles or Hopf cyclicity restricted to the center manifold are discussed.

Keywords: high-dimensional system, center manifold, Hopf bifurcation, singular point quantities

1. Introduction

This chapter is concerned with Hopf bifurcation restricted to the center manifold from the equilibrium for three-, four-, and more higher-dimensional nonlinear dynamical systems.

Let us first consider the generic real systems which take the form

$$\dot{\mathbf{x}} = A\mathbf{x} + \mathbf{f}(\mathbf{x}) \tag{1}$$

where $\mathbf{x} = (x_1, x_2, \cdots, x_n) \in \mathbb{R}^n$, $A \in \mathbb{R}^{n\times n}$, $n \in \mathbb{N}$, and $\mathbf{f}(\mathbf{x})$ is sufficiently smooth with $\mathbf{f}(\mathbf{0}) = \mathbf{0}$, $D\mathbf{f}(\mathbf{0}) = \mathbf{0}$. Then the origin is an equilibrium. For dynamical analysis of systems (1), it is very important to discuss the asymptotic behavior and the existence of periodic orbits at the origin. When the Jacobi matrix A has an eigenvalue with zero real part, the phase portraits in the vicinity of the origin is not easy to be determined. In particular, a system (1) has the following form

$$\begin{cases} \dot{\mathbf{x}}_1 = A_1\mathbf{x}_1 + \mathbf{f}_1(\mathbf{x}_1,\mathbf{x}_2) \\ \dot{\mathbf{x}}_2 = A_2\mathbf{x}_2 + \mathbf{f}_2(\mathbf{x}_1,\mathbf{x}_2) \end{cases} \tag{2}$$

where $\mathbf{x}_1 = (x_1, x_2, \ldots, x_{n_c})^T \in \mathbb{R}^{n_c}$, $\mathbf{x}_2 = (x_{n_c+1}, \ldots, x_n)^T \in \mathbb{R}^{n_s}$ with $n_c + n_s = n$, A_1 and A_2 are constant matrices, and $\mathbf{f}_1(\mathbf{x}_1, \mathbf{x}_2)$, $\mathbf{f}_2(\mathbf{x}_1, \mathbf{x}_2)$ are functions with

$$\mathbf{f}_1(\mathbf{0},\mathbf{0}) = \mathbf{0},\ \mathbf{f}_2(\mathbf{0},\mathbf{0}) = \mathbf{0},\ D\mathbf{f}_1(\mathbf{0},\mathbf{0}) = \mathbf{0},\ D\mathbf{f}_2(\mathbf{0},\mathbf{0}) = \mathbf{0}$$

Suppose that A_1 has n_c critical eigenvalues (i.e., eigenvalues with Re $\lambda = 0$) and all n_s eigenvalues of A_2 satisfy Re $\lambda < 0$. According to the Center Manifold Theorem (see, e.g., [1, 2]), there exists a (local) center manifold $\mathbf{x}_2 = \mathbf{h}(\mathbf{x}_1)$ with $\mathbf{h}(\mathbf{0}) = \mathbf{0}$, $D\mathbf{h}(\mathbf{0}) = \mathbf{0}$, and system (2) is topologically equivalent near $(\mathbf{0}, \mathbf{0})$ to the system

$$\begin{cases} \dot{\mathbf{x}}_1 = A_1\mathbf{x}_1 + \mathbf{f}_1(\mathbf{x}_1, \mathbf{h}(\mathbf{x}_1)) \\ \dot{\mathbf{x}}_2 = A_2\mathbf{x}_2. \end{cases} \tag{3}$$

The first equation in Eq. (3) is called the restriction of system (2) to its center manifold at the origin. The local center manifold, which is tangent to the $(x_1, x_2, \ldots, x_{n_c})$-plane (hyperplane) at the origin and which contains all the recurrent behavior of system (2) in a neighborhood of the origin, since the second equation in (3) is linear and has exponentially decaying solutions (see, e.g., [3]). Thus, the dynamics of Eq. (2) near a nonhyperbolic equilibrium are determined by this restriction. Generally, the local center manifold is not necessarily unique, but if the origin is a center restricted to a local center manifold for system (2), then the center manifold is unique and analytic, which is presented by the Lyapunov Center Theorem proved in Ref. [4].

If A has a simple pair of purely imaginary eigenvalues $\pm\omega i$ ($\omega > 0$), system (1) undergoes a Hopf bifurcation or multiple Hopf bifurcation in a neighborhood of the origin on the local center manifold under proper perturbations of parameters. The computation of focal values (Lyapunov coefficients) plays an important role in the study of small-amplitude limit cycles appearing in these bifurcations (see [5–14] and references therein). The projection method was used for computing the first and the second focal values (see [2]), and a perturbation technique based on multiple time scales was used for computing focal values (see [15]). For a class of three-dimensional systems, the formal series method was presented with a recursive formula for computing singular point quantities (see [16]), here the theory and methodology described in Refs. [16, 17] can be applied to n-dimensional systems, where $n \geq 4$.

If A has some zero eigenvalues for system (1), the Hopf bifurcation problem at the origin on the local center manifold becomes generally more difficult in comparison to the nondegenerate case. Take the degenerate singular point with a zero linear part in planar system, for example, the investigation of Hopf bifurcation from the equilibrium has to involve detecting the monodromy and distinguishing between a center and a focus [18, 19]. For that matter, several available approaches and corresponding results can be seen in [18–25], and one can easily find that the results on the bifurcation of limit cycles are very less. Remarkably, the author of reference [26] in 2001 gave the formal series method of calculating the singular point quantities of the degenerate critical point, which made it possible to investigate multiple Hopf bifurcation of higher degree polynomial systems [27, 28]. Here we extend its application to the local center manifold of more higher-dimensional system.

2. Case of the nondegenerate singular point

In this section, we consider Hopf bifurcation from the nondegenerate origin of system (1) restricted to the center manifold, in which the Jacobian matrix A has a pair of pure imaginary eigenvalues and its other eigenvalues are all negative. As the particular case, for planar systems there exist some good computer algebra procedure to calculate the focal values (see survey article [29], monograph [30], and references therein), here the formal series method of computing singular point quantities on the local center manifold for high-dimensional system originated from the work of [31 33] in planar systems.

2.1. The formal series method of computing nondegenerate singular point quantities on center manifold

Considering the Jacobian matrix A at the origin of system (1) has a pair of purely imaginary eigenvalues and a negative one, then by certain nondegenerate transformation, the system (1) can be changed into the following system:

$$\begin{cases} \dfrac{dx}{dt} = -y + \sum\limits_{k+j+l=2}^{\infty} A_{kjl} x^k y^j u^l = X(x, y, u), \\ \dfrac{dy}{dt} = x + \sum\limits_{k+j+l=2}^{\infty} B_{kjl} x^k y^j u^l = Y(x, y, u), \\ \dfrac{du}{dt} = -d_0 u + \sum\limits_{k+j+l=2}^{\infty} \tilde{d}_{kjl} x^k y^j u^l = \tilde{U}(x, y, u) \end{cases} \tag{4}$$

where $x, y, u, A_{kjl}, B_{kjl}, \tilde{d}_{kjl} \in \mathbb{R}$ $(k,j,l \in \mathbb{N})$ and $d_0 > 0$.

Here, we recall first the calculation method of the singular point quantities on center manifold for the above real three-dimensional nonlinear dynamical systems. By means of transformation

$$z = x + y\mathbf{i}, \quad w = x - y\mathbf{i}, \quad u = u, \quad T = \mathbf{i}t, \quad \mathbf{i} = \sqrt{-1} \tag{5}$$

system (4) is also transformed into the following complex system:

$$\begin{cases} \dfrac{dz}{dT} = z + \sum\limits_{k+j+l=2}^{\infty} a_{kjl} z^k w^j u^l = Z(z, w, u), \\ \dfrac{dw}{dT} = -w - \sum\limits_{k+j+l=2}^{\infty} b_{kjl} w^k z^j u^l = -W(z, w, u), \\ \dfrac{du}{dT} = \mathrm{i} d_0 u + \sum\limits_{k+j+l=2}^{\infty} d_{kjl} z^k w^j u^l = U(z, w, u) \end{cases} \tag{6}$$

where $z, w, T, a_{kjl}, b_{kjl}, d_{kjl} \in \mathbb{C}$ $(k, j, l \in \mathbb{N})$, the systems (4) and (6) are called concomitant.

Theorem 1 (see [16]). *For system (6), using the program of term by term calculations, we can determine a formal power series:*

$$F(z,\, w,\, u) = zw + \sum_{\alpha+\beta+\gamma=3}^{\infty} c_{\alpha\beta\gamma} z^{\alpha} w^{\beta} u^{\gamma} \tag{7}$$

such that

$$\frac{dF}{dT} = \frac{\partial F}{\partial z} Z - \frac{\partial F}{\partial y} W + \frac{\partial F}{\partial u} U = \sum_{m=1}^{\infty} \mu_m (zw)^{m+1} \tag{8}$$

where $c_{110} = 1, c_{101} = c_{011} = c_{200} = c_{020} = 0, c_{kk0} = 0, k = 2, 3, \cdots$.

Definition 1. The μ_m in the expression (8) is called the mth singular point quantity at the origin on center manifold of system (6) or (4), $m = 1, 2, \cdots$.

Theorem 2 (see [16, 34]). *For the mth singular point quantity and the mth focal value at the origin on center manifold of system (4), i.e.,* μ_m *and* v_{2m+1}, $m = 1, 2, \cdots$, *we have the following relation:*

$$v_{2m+1}(2\pi) = i\pi\mu_m + i\pi \sum_{k=1}^{m-1} \xi_m^{(k)} \mu_k \tag{9}$$

where $\xi_m^{(k)} (k = 1, 2, \cdots, m-1)$ *are polynomial functions of coefficients of system (6). Usually, it is called algebraic equivalence and written as* $v_{2m+1} \sim i\pi\mu_m$.

Based on the previous work in Ref. [16], we have developed the calculation method of the focal values on the center manifold for real four-dimensional nonlinear dynamical systems in Ref. [35]. In fact, here Theorem 1 can be generalized in the n-dimensional real systems as follows

$$\begin{cases} \dfrac{dx}{dt} = -y + \text{h.o.t.} = X(x,y,\mathbf{u}), \\ \dfrac{dy}{dt} = x + \text{h.o.t.} = Y(x,y,\mathbf{u}), \\ \dfrac{du_i}{dt} = -d_i u_i + \text{h.o.t.} = \tilde{U}_i(x,y,\mathbf{u}), \quad i = 1, 2, \cdots, n-2 \end{cases} \tag{10}$$

where $\mathbf{u} = (u_1, u_2, \cdots, u_{n-2})$, h.o.t denotes the terms in $x, y, u_1, u_2, \cdots, u_{n-2}$ with orders greater than or equal to 2, and all $d_i > 0$.

By means of transformation of Eq. (5), system (10) can be transformed into the following complex system

$$\begin{cases} \dfrac{dz}{dT} = z + \sum_{k+j+1=2}^{\infty} a_{kj1} z^k w^j \mathbf{u}^1 = Z(z,w,\mathbf{u}), \\ \dfrac{dw}{dT} = -w - \sum_{k+j+l=2}^{\infty} b_{kj1} w^k z^j \mathbf{u}^1 = -W(z,w,\mathbf{u}), \\ \dfrac{du_i}{dT} = \mathbf{i} d_i\, u_i + \sum_{k+j+1=2}^{\infty} d_{kj1} z^k w^j \mathbf{u}^1 = U_i(z,w,\mathbf{u}), \quad i = 1, 2, \cdots, n-2 \end{cases} \tag{11}$$

where the subscript "$kj1$" denotes "$kjl_1 \cdots l_{n-2}$", $\mathbf{u}^1 = u_1^{l_1} u_2^{l_2} \cdots u_{n-2}^{l_{n-2}}$, and $l = \sum_{i=1}^{n-2} l_i$, all $u_i \in \mathbb{R}$, $z, w, T, a_{kj1}, b_{kj1}, d_{kj1} \in \mathbb{C}$ $(k, j, l_i \in \mathbb{N})$, we call that system (10) and system (11) are concomitant.

Theorem 3. *For system (11), using the program of term by term calculations, we can determine a formal power series:*

$$F(z,w,\boldsymbol{u}) = zw + \sum_{\alpha+\beta+\ell=3}^{\infty} c_{\alpha\beta\,\ell}\, z^{\alpha} w^{\beta} \boldsymbol{u}^{\ell} \tag{12}$$

such that

$$\frac{dF}{dT} = \frac{\partial F}{\partial z} Z - \frac{\partial F}{\partial y} W + \sum_{i=1}^{n-2} \frac{\partial F}{\partial u_i} U_i = \sum_{m=1}^{\infty} \mu_m (zw)^{m+1} \tag{13}$$

where the subscript "$\alpha\beta\ell$" denotes "$\alpha\beta\gamma_1\cdots\gamma_{n-2}$", $\boldsymbol{u}^{\ell} = u_1^{\gamma_1} u_2^{\gamma_2} \cdots u_{n-2}^{\gamma_{n-2}}$, and $\ell = \sum_{i=1}^{n-2} \gamma_i$, and more setting $c_{\alpha\beta\ell} = 0$ with $0 \leq \alpha + \beta + \ell \leq 2$ except for $c_{110} = 1$, and $c_{kk0} = 0$ with $k \geq 2$.

Proof. It is very similar to the proving course of Theorem 1.3.1 in [16], by computing carefully and comparing the above power series with the two sides of (13), we can obtain the expression of μ_m.

Definition 2. The μ_m in the expression (13) is called the mth singular point quantity at the origin on center manifold of system (11) or (10), $m = 1, 2, \cdots$.

Remark 1. Similar to Theorem 2, there exists a equivalence between μ_m and v_{2m+1}, namely, if $\mu_1 = \mu_2 = \cdots = \mu_{m-1} = 0, \mu_m \neq 0$, then $v_3 = v_5 = \cdots = v_{2m-1} = 0, v_{2m+1} = i\pi\mu_m$, $m = 1, 2, \cdots$, and vice versa.

Corollary 1. *The origin of system (10) or (11) is a center restricted to the center manifold if and only if $\mu_m = 0$ for all m.*

Remark 2. From the relation given by Remark 1 and Corollary 1, the center-focus problem and Hopf bifurcation of equilibrium point restricted to the center manifold can be figured out by the calculation of singular point quantities for system (10).

2.2. An example of four-dimensional system

Recently, the study of chaos has become a hot research topic, and the attention of many researchers is turning to 4D systems from 3D dynamical systems, for example, the authors of Ref. [36] investigated Hopf bifurcation of a 4D-hyoerchaotic system by applying the normal form theory in 2012, but its multiple Hopf bifurcation on the center manifold have not been considered. Here, we will investigate the system further by computing the singular point quantities of its equilibrium point, which takes the following form

$$\begin{cases} \dot{x}_1 = a(x_2 - x_1) \\ \dot{x}_2 = cx_1 - x_2 + x_4 - x_1 x_3 \\ \dot{x}_3 = x_1 x_2 - bx_3 + ex_1^2 \\ \dot{x}_4 = -Kx_2 \end{cases} \tag{14}$$

where $a, b, c, e, K \in \mathbb{R}$. Obviously, system (14) has only one isolated equilibrium: $O(0,0,0,0)$ when $K \neq 0$. Therefore, we only need to consider O. The Jacobian matrix of system (14) at O is

$$A = \begin{pmatrix} -a & a & 0 & 0 \\ c & -1 & 0 & 1 \\ 0 & 0 & -b & 0 \\ 0 & -K & 0 & 0 \end{pmatrix}$$

with the characteristic equation:

$$(\lambda + b)(\lambda^3 + (a+1)\lambda^2 + (a - ac + K)\lambda + aK = 0. \tag{15}$$

To guarantee that A has a pair of purely imaginary eigenvalues $\pm i\,\omega(\omega > 0)$ and two negative real eigenvalues λ_1, λ_2, we let its characteristic equation take the form

$$(\lambda^2 + \omega^2)(\lambda - \lambda_1)(\lambda - \lambda_2) = 0.$$

Thus, we obtain the critical condition of Hopf bifurcation at O:

$$a^2(c-1) = \omega^2,\ K = a(a+1)(c-1),\ \ \lambda_1 = -b,\ \lambda_2 = -a-1 \tag{16}$$

where $a > -1$, $b > 0$, $c > 1$, namely, $c = \frac{a^2+\omega^2}{a^2}$, $K = \frac{(a+1)\omega^2}{a}$. Under the conditions (16), one can find a nondegenerate matrix

$$P = \begin{pmatrix} -\dfrac{\mathbf{i}a^2}{(a+1)(a+i\omega)\omega} & \dfrac{\mathbf{i}a^2}{(a+1)(a-i\omega)\omega} & 0 & -\dfrac{a^2}{\omega^2} \\ -\dfrac{\mathbf{i}a}{(a+1)\omega} & \dfrac{\mathbf{i}a}{a\omega+\omega} & 0 & \dfrac{a}{\omega^2} \\ 0 & 0 & 1 & 0 \\ 1 & 1 & 0 & 1 \end{pmatrix}$$

such that

$$P^{-1}AP = \begin{pmatrix} \omega\mathbf{i} & 0 & 0 & 0 \\ 0 & -\omega\mathbf{i} & 0 & 0 \\ 0 & 0 & -b & 0 \\ 0 & 0 & 0 & -a-1 \end{pmatrix} \tag{17}$$

Namely, we can use the nondegenerate transformation and the time rescaling: $T = it\omega$ to make the system (14) become the following same form as the complex system (11) with $n = 4$:

$$\begin{cases} \dfrac{\mathrm{d}z}{\mathrm{d}T} = z + \sum_{k+j+l+n=2}^{2} a_{kjln} z^k w^j u^l v^n = Z(z,w,u,v), \\ \dfrac{\mathrm{d}w}{\mathrm{d}T} = -w - \sum_{k+j+l+n=2}^{2} b_{kjln} w^k z^j u^l v^n = -W(z,w,u,v), \\ \dfrac{\mathrm{d}u}{\mathrm{d}T} = \dfrac{b\mathbf{i}}{\omega} u + \sum_{k+j+l+n=2}^{2} d_{kjln} z^k w^j u^l v^n = U(z,w,u,v), \\ \dfrac{\mathrm{d}v}{\mathrm{d}T} = \dfrac{(a+1)\mathbf{i}}{\omega} v + \sum_{k+j+l+n=2}^{2} e_{kjln} z^k w^j u^l v^n = V(z,w,u,v) \end{cases} \tag{18}$$

where $u \in \mathbb{R}$, z, w, $T \in \mathbb{C}$, and all $a_{kjln} = b_{kjln} = d_{kjln} = e_{kjln} = 0$ except the following coefficients

$$a_{0011} = \frac{a^3 + a^2(1+\mathbf{i}\omega) + \mathbf{i}a\omega}{2\omega^2(a+\mathbf{i}\omega+1)}, \quad a_{0110} = \frac{a(\omega - \mathbf{i}a)}{2\omega\left(a^2 + a + \omega(\omega - \mathbf{i})\right)},$$

$$b_{kjln} = \bar{a}_{kjln} \ (ikjl = 0011, 0110),$$

$$d_{0002} = \frac{\mathbf{i}a^3(1-a)e}{\omega^5}, \quad d_{0101} = -\frac{a^4(2e+1) - a^3(1+\mathbf{i}\omega)}{(a+1)\omega^4(a-\mathbf{i}\omega)},$$

$$d_{0200} = \frac{a^3\omega + \mathbf{i}a^4(e+1)}{(a+1)^2\omega^3(a-\mathbf{i}\omega)^2}, \quad d_{1001} = \frac{a^4(2e+1) + a^3(\mathbf{i}\omega - 1)}{(a+1)\omega^4(a+\mathbf{i}\omega)},$$

$$d_{1100} = -\frac{2\mathbf{i}a^4(e+1)}{(a+1)^2\omega^3(a^2+\omega^2)}, \quad d_{2000} = -\frac{a^3\omega - \mathbf{i}a^4(e+1)}{(a+1)^2\omega^3(a+\mathbf{i}\omega)^2},$$

$$e_{0011} = -\frac{\mathbf{i}a(a+1)}{\omega(a^2+2a+\omega^2+1)}, \quad e_{0110} = -\frac{a}{(a-\mathbf{i}\omega)(a^2+2a+\omega^2+1)},$$

$$e_{1010} = \frac{a}{(a+\mathbf{i}\omega)(a^2+2a+\omega^2+1)}$$

where $\bar{a}_{kjln}$ denotes the conjugate complex number of a_{kjln}.

According to Theorem 3, we obtain the recursive formulas of $c_{\alpha\beta\gamma}$ and μ_m.

Theorem 5. *For system (18), setting $c_{\alpha\beta\gamma\lambda} = 0$ with $0 \le \alpha+\beta+\gamma+\lambda \le 2$ except for $c_{1100} = 1$, and $c_{kk00} = 0$ with $k \ge 2$, we can derive successively and uniquely the terms of the following formal series (12) with $n = 4$, such that (13) with $n = 4$ holds and if $\alpha \ne \beta$ or $\alpha = \beta$, $\lambda^2 + \gamma^2 \ne 0$, $c_{\alpha\beta\gamma\lambda}$ is determined by following recursive formula:*

$$\begin{aligned} c_{\alpha\beta\gamma\lambda} = {} & \frac{\omega}{\omega(\alpha-\beta) + i(b\,\gamma + (a+1)\,\lambda)} \\ & \{-d_{2000}(1+\gamma)c[\alpha-2,\beta,\gamma+1,\lambda] - d_{1100}(\gamma+1)c[\alpha-1,\beta-1,\gamma+1,\lambda] - \\ & e_{1010}(\lambda+1)c[\alpha-1,\beta,\gamma-1,\lambda+1] - d_{1001}(\gamma+1)c[\alpha-1,\beta,\gamma+1,\lambda-1] + \\ & b_{0110}(\beta+1)c[\alpha-1,\beta+1,\gamma-1,\lambda] - d_{0200}(\gamma+1)c[\alpha,\beta-2,\gamma+1,\lambda] - \\ & e_{0110}(\lambda+1)c[\alpha,\beta-1,\gamma-1,\lambda+1] - d_{0101}(\gamma+1)c[\alpha,\beta-1,\gamma+1,\lambda-1] - \\ & e_{0011}\lambda c[\alpha,\beta,\gamma-1,\lambda] - d_{0002}(\gamma+1)c[\alpha,\beta,\gamma+1,\lambda-2] + \\ & b_{0011}(\beta+1)c[\alpha,\beta+1,\gamma-1,\lambda-1] - a_{0110}(\alpha+1)c[\alpha+1,\beta-1,\gamma-1,\lambda] - \\ & a_{0011}(\alpha+1)c[\alpha+1,\beta,\gamma-1,\lambda-1]\} \end{aligned} \tag{19}$$

and for any positive integer m, μ_m is determined by following recursive formula:

$$\begin{aligned} \mu_m = {} & d_{2000}c[-2+m, m, 1, 0] \\ & + d_{1100}c[-1+m, -1+m, 1, 0] + d_{0200}c[m, -2+m, 1, 0] \end{aligned} \tag{20}$$

and when $\alpha < 0$ or $\beta < 0$ or $\gamma < 0$ or $\lambda < 0$ or $\alpha = \beta$, $\gamma = \lambda = 0$, we have let $c_{\alpha\beta\gamma\lambda} = 0$, and where each $c[\alpha,\beta,\gamma,\lambda]$ denotes $c_{\alpha\beta\gamma\lambda}$.

By applying the above formulas in the Mathematica symbolic computation system, we figure out easily the first two singular point quantities of the origin of system (18):

$$\begin{aligned} \mu_1 &= iaf_1\,[|a|\,b\,c\,(a+1)^2 d_0]^{-1}, \\ \mu_2 &= 108ia^3b^4 f_2 f_3^2 f_4\,[|a|\,c^2 d_0 d_1^{\,2} d_2^{\,4} d_3]^{-1} \end{aligned} \tag{21}$$

where

$$
\begin{aligned}
f_1 &= 8a^3ce + 8a^3c - 8a^3e - 8a^3 - 2a^2bce + 2a^2be + 8a^2ce + 8a^2c \\
&\quad - 8a^2e - 8a^2 + ab^2c + 3ab^2e + 2ab^2 + 2abc - 2ab + 3b^2e + 3b^2, \\
f_2 &= (2a + b + 2)^3(2ae + 2a - b)(e + 1), \\
f_3 &= 4a^2e + 4a^2 - 3abe - 2ab + 4ae + 4a + b, \\
f_4 &= 8a^5c^2 - 16a^5c + 8a^5 - 2a^4bc^2 + 2a^4bc + 8a^4c^2 - 16a^4c + 8a^4 + 2a^3b^2c \\
&\quad - 2a^3b^2 - 4a^3bc + 4a^3b - 5a^2b^3c + 4a^2b^3 + 2a^2b^2c \\
&\quad - 2a^2b^2 - 2a^2bc + 2a^2b - 2ab^3 - b^3, \\
d_0 &= (a^2c + 2a + 1)(4a^2c - 4a^2 + b^2)(c-1)^{3/2}, \\
d_1 &= 8a^3c - 8a^3 - 2a^2bc + 2a^2b + 8a^2c - 8a^2 + 3ab^2 + 3b^2, \\
d_2 &= 8a^2e + 8a^2 - 2abe + 8ae + 8a + b^2 + 2b, \\
d_3 &= 9a^2c - 8a^2 + 2a + 1,
\end{aligned}
$$

and the above expression of μ_2 is obtained under the condition of $\mu_1 = 0$.

From Remark 1 and the singular point quantities (21), we have

Theorem 6. *For the flow on center manifold of the system (14), the first 2 focal values of the origin are as follow*

$$
v_3 = i\pi\mu_1, \qquad v_5 = i\pi\mu_2 \tag{22}
$$

where the expression of v_5 is obtained under the condition of $v_3 = 0$.

Remark 3. In contrast to the result and process in [36], one can easily see that our first quantity is basically consistent with its characteristic exponent of bifurcating periodic solutions, and our algorithm is easy to realize with computer algebra system due to the linear recursion formulas, and more convenient to investigate the multiple Hopf bifurcation on center manifold.

Considering its Hopf bifurcation form of Theorem 6, we have the following:

Theorem 7. *At least two small limit cycles can be bifurcated from the origin of the 4D-hyoerchaotic system (14), which lie in the neighborhood of the origin restricted to the center manifold.*

The rigorous proof of the above theorem is very similar to the previous ones in [14, 16], namely, by calculating the Jacobian determinant with respect to the functions v_3, v_5 and its variables, which will not be given here.

3. Case of the degenerate singular point

Up till now, study on bifurcation of limit cycles from the degenerate singularity of higher dimensional nonlinear systems (1) is hardly seen in published references. Here, we will investigate the Hopf bifurcation problem from the high-order critical point on the center manifold.

3.1. The formal series method of computing degenerate singular point quantities on center manifold

Let us consider the real n-dimensional systems with two zero eigenvalues and zero linear part as follows

$$\begin{cases} \dfrac{dx}{dt} = (\delta x - y)(x^2+y^2)^q + \sum\limits_{k+j+1=2q+2}^{\infty} A_{kj1}x^k y^j \mathbf{u}^1 = X(x,y,\mathbf{u}), \\ \dfrac{dy}{dt} = (x - \delta y)(x^2+y^2)^q + \sum\limits_{k+j+1=2q+2}^{\infty} B_{kj1}x^k y^j \mathbf{u}^1 = Y(x,y,\mathbf{u}), \\ \dfrac{du_i}{dt} = -d_i u_i + \sum\limits_{k+j+1=2}^{\infty} d_{kjl} z^k w^j \mathbf{u}^1 = U_i(x,y,\mathbf{u}), \quad i = 1,2,\cdots,n-2 \end{cases} \tag{23}$$

where the subscript "$kj1$" denotes "$kjl_1 \cdots l_{n-2}$", $\mathbf{u}^1 = u_1^{l_1}u_2^{l_2}\cdots u_{n-2}^{l_{n-2}}$, and $l = \sum\limits_{i=1}^{n-2} l_i$, all $d_i > 0$, x, y, u_i, t, δ, A_{kjl}, B_{kjl}, $d_{kjl} \in \mathbb{R}$, q, k, j, $l_i \in \mathbb{N}$. Obviously, the origin of system (23) is a high-order degenerate singular point with two zero eigenvalues and $n-2$ negative ones.

In order to discuss the calculation method of the focal values on center manifold of the system (23), from the center manifold theorem [1], we take an approximation to the center manifold:

$$\mathbf{u} = \mathbf{u}(x,y) = \mathbf{u}_2(x,y) + \mathbf{h.o.t.} \tag{24}$$

where $u = (x_1, x_2, \cdots, x_{n-2})^T$, $\mathbf{u}_2$ is a quadratic homogeneous polynomial vector in x and y, and $\mathbf{h.o.t.}$ denotes the terms with orders greater than or equal to 3. Substituting $\mathbf{u} = \mathbf{u}(x,y)$ into the equations of system (23), we obtain a real planar polynomial differential system as follows

$$\begin{cases} \dfrac{dx}{dt} = (\delta x - y)(x^2+y^2)^q + \sum\limits_{k=2q+2}^{\infty} X_k(x,y) = \tilde{X}(x,y), \\ \dfrac{dy}{dt} = (x - \delta y)(x^2+y^2)^q + \sum\limits_{k=2q+2}^{\infty} Y_k(x,y) = \tilde{Y}(x,y) \end{cases} \tag{25}$$

where $X_k(x,y)$, $Y_k(x,y)$ are homogeneous polynomials of degree k, and the origin is degenerate with a zero linear part.

For system (25), some significant works have been done in Refs. [26] and [27]. Let us recall the related notions and results.

By means of transformation (5)

$$z = x + y\mathbf{i}, \quad w = x - y\mathbf{i}, \quad u = u, \quad T = \mathbf{i}\,t, \quad \mathbf{i} = \sqrt{-1},$$

system (25) is transformed into following system:

$$\begin{cases} \dfrac{dz}{dT} = (1 - \mathbf{i}\delta) z^{q+1} w^q + \sum\limits_{k+j=2q+2}^{\infty} a_{kj} z^k w^j = Z(z,w), \\ \dfrac{dw}{dT} = -(1 + \mathbf{i}\delta) z^q w^{q+1} - \sum\limits_{k+j=2q+2}^{\infty} b_{kj} z^k w^j = -W(z,w) \end{cases} \tag{26}$$

where z, w, T are complex variables and for any positive integer k, j, we have $a_{kj} = \overline{b}_{kj}$, then systems (25) and (26) are called concomitant.

For any positive integer k, we denote

$$f_k(z,w) = \sum_{\alpha+\beta=k} c_{\alpha\beta} z^\alpha w^\beta$$

a homogeneous polynomial of degree k with $c_{00} = 1$, $c_{kk} = 0$, $k = 1, 2,\cdots$.

Theorem 8 ([26, 27]). *For system (26) with $\delta = 0$, we can derive successively the terms of the following formal series:*

$$F(z,w) = zw\left[1 + \sum_{m=1}^{\infty} \frac{f_{m(2q+3)}(z,w)}{(zw)^{m(q+1)}}\right] \tag{27}$$

such that

$$\frac{dF}{dT} = \frac{\partial F}{\partial z}Z - \frac{\partial F}{\partial w}W = (zw)^q \sum_{m=1}^{\infty} \mu_m (zw)^{m+1}. \tag{28}$$

Definition 3. If $\delta = 0$ holds, μ_m in expression (28) is called the mth singular point quantity at the degenerate singular point for system (26) or (1.3.26) is also called the mth singular point quantity of the origin on the center manifold of system (23), where $m = 1, 2, \cdots$.

Similar to Theorem 2, there also exists a equivalence between the mth singular point quantity and the mth focal value $v_{2m+1}(2\pi)$ at the origin on center manifold of system (23).

Theorem 9. *For system (23) with $\delta = 0$, and any positive integer m, the following assertion holds: $v_{2m+1}(2\pi)\tilde{i}\pi\mu_m$, namely*

$$v_{2m+1}(2\pi) = i\pi\left(\mu_m + \sum_{k=1}^{m-1} \xi_m^{(k)} \mu_k\right), \tag{29}$$

where $\xi_m^{(k)}$ $(k = 1, 2, \cdots, m-1)$ are polynomial functions of coefficients of system (26). Then, the relation between $v_{2m+1}(2\pi)$ and μ_m is called the algebraic equivalence.

Remark 4. In fact, from Theorem 2, for any positive integer $m = 2, 3, \cdots$, if $\mu_1 = \mu_2 = \cdots = \mu_{m-1} = 0$ and $v_1(2\pi) = v_3(2\pi) \cdots = v_{2m-1}(2\pi) = 0$ hold, and vice versa. And more the stability and bifurcation of the origin of system (23) can be figured out by calculating the singular point quantities.

Corollary 2. *The origin of system (23) is a center restricted to the center manifold if and only if $\mu_m = 0$ for all m.*

3.2. An example of three-dimensional system

Now we consider an example for system (23) with $n = 3$, it can be put in its concomitant form as follows

$$\begin{cases} \frac{dz}{dT} = (1-\mathbf{i}\delta)z^2w + uz\,(a_{20}z^2 + a_{11}zw + a_{02}w^2) = Z, \\ \frac{dw}{dT} = -(1+\mathbf{i}\delta)zw^2 - uw\,(b_{20}w^2 + b_{11}wz + b_{02}z^2) = -W, \\ \frac{du}{dT} = \mathbf{i}u + \mathbf{i}d_1zw = U, \end{cases} \tag{30}$$

where $d_1 \neq 0$ and

$$a_{ij} = A_i + \mathbf{i}B_i,\ b_{ij} = A_i - \mathbf{i}B_i,\ A_i, B_i \in \mathbb{R},\ i,j = 0,1,2, \tag{31}$$

namely, $a_{ij} = \overline{b}_{ij}$. Then for the center manifold of system (30), from the transformation (5), we can determine the formal expression (24): $u = u(x,y) = \tilde{u}(z,w)$, thus obtain

$$\begin{cases} \frac{dz}{dT} = (1-\mathbf{i}\delta)z^2w + \tilde{u}z\,(a_{20}z^2 + a_{11}zw + a_{02}w^2) = \tilde{Z}, \\ \frac{dw}{dT} = -(1+\mathbf{i}\delta)zw^2 - \tilde{u}w\,(b_{20}w^2 + b_{11}wz + b_{02}z^2) = -\tilde{W} \end{cases}. \tag{32}$$

Remark 5. For system (32), the corresponding $n = 1$ in (27) and (28) of **Theorem 8**, we figure out that each μ_m is related to only the coefficients of the first $2m+3$ order terms of system (32), $m = 1, 2, \cdots$. Here, we determine the above $\tilde{u}$ just to the sixth-order term as follows

$$\tilde{u}(z,w) = \sum_{k=2}^{6} \tilde{u}_k(z,w) \tag{33}$$

where $\tilde{u}_k$ is a homogeneous polynomial in z,w of degree k and

$$\begin{aligned} \tilde{u}_2 &= -d_1zw,\ \tilde{u}_4 = 2\delta d_1z^2w^2,\ \tilde{u}_3 = \tilde{u}_4 = \tilde{u}_5 = 0, \\ \tilde{u}_6 &= -\mathbf{i}d_1wz((a_{02}-b_{20})d_1w^3z + (a_{11}d_1 - b_{11}d_1 - 8\mathbf{i}\delta^2)w^2z^2 \\ &\quad + (a_{20}-b_{02})d_1wz^3). \end{aligned} \tag{34}$$

Hence, $\tilde{Z}$ and $\tilde{W}$ in system (32) are two polynomials with degree 9.

Theorem 10. *For system (32) with $\delta = 0$, we can derive successively the terms of the formal series (27), such that (28) holds ($c_{\alpha\beta}$, μ_m in Appendix A).*

Applying the powerful symbolic computation function of the Mathematica system and the recursive formulas in Theorem 10, and from Remark 5, we obtain the first three singular point quantities as follows

$$\begin{aligned} \mu_1 &= -d_1(a_{11}-b_{11}), \\ \mu_2 &= d_1^2(b_{20}b_{02} - a_{20}a_{02}), \\ \mu_3 &= -2\mathbf{i}d_1^2(a_{02}a_{20} + b_{02}b_{20} - a_{02}b_{02} - a_{20}b_{20}) \end{aligned} \tag{35}$$

In the above expression of each μ_k, $k = 2, 3$, we have already let $\mu_1 = \cdots = \mu_{k-1} = 0$.

Thus, from Theorem 9 and Eqs. (35) and (31), we have

Theorem 11. *For the flow on center manifold of system* $(30)_{,\delta=0}$, *the first three focal values* $v_{2i+1}(2\pi)$ $(i=1,2,3)$ *of the origin are as follows*

$$\begin{aligned} v_3 &= 2\pi d_1 B_1, \\ v_5 &= 2\pi d_1^2\,(A_2B_0 + A_0B_2), \\ v_7 &= 2\pi d_1^2\,[(A_0 - A_2)^2 + (B_0 + B_2)^2] \end{aligned} \tag{36}$$

Theorem 12. *For the flow on center manifold of* $(30)_{\delta=0}$, *the origin is a three-order weak focus, i.e.,* $v_3 = v_5 = 0$, $v_7 \neq 0$ *if and only if*

$$B_1 = 0,\ A_2B_0 + A_0B_2 = 0 \text{ and } (A_0 - A_2)^2 + (B_0 + B_2)^2 \neq 0 \tag{37}$$

Remark 6. For the coefficients of system $(30)_{\delta=0}$, there exists necessarily a group of critical values: $A_i = A_i^*$, $B_i = B_i^*$ $(i=0,1,2)$ such that the conditions (37) hold, for example:

$$A_1^* = B_1^* = 0,\ A_0^* = B_0^* = 1,\ B_2^* = -A_2^* = 13 \tag{38}$$

Now we consider Hopf bifurcation of limit cycles from the origin for perturbed system (30).

Theorem 13. *At least three limit cycles can be bifurcated from the origin of system (30) restricted to the center manifold, which lie in the neighborhood of the origin.*

Proof. From Theorem 11, one can easily calculate the Jacobian determinant with respect to the functions v_3,v_5,v_7 and variables B_1,B_0,A_0,

$$J = \frac{\partial(v_3, v_5, v_7)}{\partial(B_1, B_0, A_0)} = -2\pi^3 d_1^5[8(A_0A_2 - A_2^2 - B_0B_2 - B_2^2)] \tag{39}$$

Considering the conditions (37) of Theorem 12 and substituting the group of critical values of Eq. (38) into Eq. (39), we obtain $J = 649\pi^3 d_1^5 \neq 0$. Thus, we take some appropriate perturbations for the coefficients of system (32) to make the following two conditions:

$$(v_1(2\pi) - 1)v_3 < 0,\ v_3v_5 < 0,\ v_5v_7 < 0 \tag{40}$$

and

$$|e^{2\pi\delta} - 1| \ll |v_3| \ll |v_5| \ll |v_7| \tag{41}$$

hold, one must obtain that the succession function on the center manifold has three small real positive roots, just the system (30) has at least three limit cycles in the neighborhood of the origin. We can refer to references [16, 26, 27] for more details about the construction of limit cycles.

Remark 7. In general, in order to find more limit cycles in the neighborhood of the origin of system (30), we should add more higher order terms of $\tilde{u}(z,w)$ determined in Eq. (33). Here we propose a conjecture that system (30) has at most three limit cycles through Hopf bifurcation restricted to a center manifold from the origin. However, the center conditions or integrability at the degenerate singularity will need further study.

4. Conclusion and discussion

The two classes of methods for computing the nondegenerate and degenerate singular point quantities on center manifold of the three-, four-, and more higher dimensional polynomial systems are discussed here, and more as the applications of them, the multiple limit cycles or Hopf cyclicity of two typical nonlinear dynamic systems restricted to the corresponding center manifolds are investigated.

Appendix A

$$
\begin{aligned}
& c[\alpha,\beta] = \\
& \frac{1}{5(\alpha-\beta)} d_1\{b_{02}^2(3\beta-2\alpha) + a_{20}b_{02}(20-\beta-\alpha)-a_{20}^2(20+2\beta-3\alpha)) \\
& \times d_1 c[\alpha-17,\beta-13] + ((a_{11}b_{02}+a_{20}b_{11})(20-\beta-\alpha)-2b_{02}b_{11}(5-3\beta+2\alpha)- \\
& 2a_{11}a_{20}(15+2\beta-3\alpha))d_1 c[\alpha-16,\beta-14] + ((a_{02}b_{02}+a_{11}b_{11}+a_{20}b_{20})(20- \\
& \beta-\alpha)-(a_{11}^2+2a_{02}a_{20})(10+2\beta-3\alpha)-(b_{11}^2+2b_{02}b_{20})(10-3\beta+2\alpha))d_1 c[\alpha- \\
& 15,\beta-15] + ((a_{02}b_{11}+a_{11}b_{20})(20-\beta-\alpha)-2b_{11}b_{20}(15-3\beta+2\alpha)-2a_{02}a_{11}(5+ \\
& 2\beta-3\alpha))d_1 c[\alpha-14,\beta-16] + (a_{02}b_{20}(20-\beta-\alpha)-b_{20}^2(20-3\beta+2\alpha)- \\
& a_{02}^2(2\beta-3\alpha))d_1 c[\alpha-13,\beta-17]-b_{02}(5+3\beta-2\alpha)+a_{20}(5+2\beta-3\alpha))\mathbf{i}c[\alpha- \\
& 6,\beta-4]-(b_{11}(3\beta-2\alpha)+a_{11}(2\beta-3\alpha))\mathbf{i}\, c[\alpha-5,\beta-5] \\
& +(b_{20}(5-3\beta+2\alpha)+a_{02}(5-2\beta+3\alpha))\mathbf{i}c[\alpha-4,\beta-6]
\end{aligned}
$$

$$
\begin{aligned}
& \tilde{\mu}\,[\alpha] = -\frac{d_1}{5}\{(a_{20}^2(\alpha-20)+2a_{20}b_{02}(10-\alpha)+b_{02}^2\alpha)d_1 c[\alpha-17,\alpha-13] \\
& +(2a_{11}a_{20}(\alpha-15)-2(a_{11}b_{02}+a_{20}b_{11})(\alpha-10)+2b_{02}b_{11}(\alpha-5))d_1 c[\alpha- \\
& 16,\alpha-14] + ((a_{11}^2+2a_{02}a_{20}-2a_{02}b_{02}-2a_{11}b_{11}+b_{11}^2-2a_{20}b_{20}+2b_{02}b_{20})(\alpha- \\
& 10))d_1 c[\alpha-15,\alpha-15] + 2((a_{02}b_{11}+a_{11}b_{20})(10-\alpha)-b_{11}b_{20}(15-\alpha)-a_{02}a_{11}(5- \\
& \alpha))d_1 c[\alpha-14,\alpha-16] + (b_{20}^2(\alpha-20)-2(a_{02}b_{20})(\alpha-10)+a_{02}^2\alpha)d_1 c[\alpha-13,\alpha- \\
& 17] + (a_{20}(\alpha-5)-b_{02}(5+\alpha))\mathbf{i}\, c[\alpha-6,\alpha-4] + (a_{11}-b_{11})\alpha\mathbf{i}i\, c[\alpha-5,\alpha- \\
& 5]-(b_{20}(\alpha-5)-a_{02}(5+\alpha))i\, c[\alpha-4,\alpha-6]\}, \\
& \mu_m = \tilde{\mu}\,[5m],
\end{aligned}
$$

where $c[k,j] = c_{kj}$.

Acknowledgements

This work was supported by Natural Science Foundation of China grants (11461021, 11261013), Nature Science Foundation of Guangxi (2015GXNSFAA139011), Research Foundation of Hezhou University (No.HZUBS201302), and Guangxi Education Department Key Laboratory of Symbolic Computation and Engineering Data Processing.

MR(2000) Subject Classification: 34C05, 37C07

Author details

Qinlong Wang[1*], Bo Sang[2] and Wentao Huang[1]

*Address all correspondence to: wqinlong@163.com

1 School of Science, Hezhou University, Hezhou, P.R. China

2 School of Mathematical Sciences, Liaocheng University, Liaocheng, P.R. China

References

[1] J. Carr, Applications of Centre Manifold Theory, Appl. Math. Sci., Vol. 35, Springer, New York, 1981.

[2] Y. A. Kuznetsov, Elements of Applied Bifurcation Theory, Springer-Verlag, New York, 2004.

[3] J. Sijbrand, Properties of center manifolds, Trans. Am. Soc., 1985, 289: 431–469.

[4] Y.N. Bibikov, Local theory of nonlinear analytic ordinary differential equations, in: Lecture Notes in Mathematics, Vol. 702, Springer-Verlag, New York, 1979.

[5] L. Barreira, C. Valls, J. Llibre, Integrability and limit cycles of the Moon-Rand system, Int. J. Nonlin. Mech., 2015, 69: 129–136.

[6] M. Gyllenberg, P. Yan, Four limit cycles for a three-dimensional competitive Lotka-Volterra system with a heteroclinic cycle, Comput. Math. Appl., 2009, 58: 649–669.

[7] M.A. Han, P. Yu, Ten limit cycles around a center-type singular point in a 3-d quadratic system with quadratic perturbation, Appl. Math. Lett., 2015, 44: 17–20.

[8] J. Llibre, C. Valls, Hopf bifurcation of a generalized Moon-Rand system, Commun. Nonlinear Sci. Numer. Simulat., 2015, 20: 1070–1077.

[9] A. Mahdi, V.G. Romanovski, D.S. Shafer, Stability and periodic oscillations in the Moon-Rand systems, Nonlinear. Anal. Real, 2013, 14: 294–313.

[10] Y. Tian, P. Yu, Seven limit cycles around a focus point in a simple three-dimensional quadratic vector field, Int. J. Bifurcation Chaos, 2014, 24: 1450083–1450092.

[11] P. Yu, M.A. Han, Eight limit cycles around a center in quadratic Hamiltonian system with third-order perturbation, Int. J. Bifurcation Chaos, 2013, 23: 1350005–1350022.

[12] J. Hofbauer, J.W.-H. So, Multiple limit cycles for three dimensional Lotka-Volterra Equations, Appl. Math. Lett., 1994, 7: 65–70.

[13] Z. Lu, Y. Luo, Two limit cycles in three-dimensional Lotka-Volterra systems, Comput. Math. Appl., 2002, 44: 51–66.

[14] Q. Wang, W. Huang, B.L. Li, Limit cycles and singular point quantities for a 3D Lotka-Volterra system, Appl. Math. Comput., 2011, 217: 8856–8859.

[15] P. Yu, Computation of normal forms via a perturbation technique, J. Sound Vibration, 1998, 211: 19–38.

[16] Q. Wang, Y. Liu, H. Chen, Hopf bifurcation for a class of three-dimensional nonlinear dynamic systems, Bull. Sci. Math., 2010, 134: 786–798.

[17] B. Sang, Center problem for a class of degenerate quartic systems, Electron. J. Qual. Theory Diff. Equation, 2014, 74(2014): 1–17.

[18] A. Gasull, J. Llibre, V. Manosa, F. Manosas, The focus-centre problem for a type of degenerate system, Nonlinearity, 2000, 13: 699–729.

[19] I.A. García, S. Maza, A new approach to center conditions for simple analytic monodromic singularities, J. Diff. Equations, 2010, 248: 363–380.

[20] A. Gasull, V. Manosa, F. Manosas, Monodromy and stability of a class of degenerate planar critical points, J. Diff. Equations, 2002, 182: 169–190.

[21] V. Mañosa, On the center problem for degenerate singular points of planar vector fields, Intl. J. Bifurcation Chaos Appl. Sci. Eng., 2002, 12: 687–707.

[22] I.A. García, J. Giné, M. Grau, A necessary condition in the monodromy problem for analytic differential equations on the plane, J. Symbolic. Comput., 2006, 41: 943–958.

[23] A. Algaba, E. Freire, E. Gamerob, C. García, Monodromy, center-focus and integrability problems for quasi-homogeneous polynomial systems, Nonlinear Anal., 2010, 72: 1726–1736.

[24] P. De Jager, An example of a quadratic system with a bounded separatrix cycle having at least two limit cycles in its interior, Delft Prog. Rep., 1986C1987, 11: 141–150.

[25] I.A. García, H. Giacomini, M. Grau, Generalized Hopf bifurcation for planar vector fields via the inverse integrating factor, J. Dyn. Diff. Equation, 2011, 23: 251–281.

[26] Y. Liu, Theory of center-focus for a class of higher-degree critical points and infinite points, Sci. China (Series A), 2001, 44: 37–48.

[27] H. Chen, Y. Liu, X. Zeng, Center conditions and bifurcation of limit cycles at degenerate singular points in a quintic polynomial differential system, Bull. Sci. Math., 2005, 129: 127–138.

[28] W. Huang, Y. Liu, A polynomial differential system with nine limit cycles at infinity. Comput. Math. Appl., 2004, 48: 577–588.

[29] J. Li, Hilberts 16th problem and bifurcations of planar polynomial vector fields, Intl. J. Bifurcation Chaos, 2003, 33: 47–106.

[30] M. Han, Bifurcation Theory of Limit Cycles, Mathematics Monograph Series, Vol. 25, Science Press, Beijing, 2013.

[31] Y. Liu, J. Li, Theory of values of singular point in complex autonomous differential system, Sci. China (Series A), 1990, 33: 10–24.

[32] H. Chen, Y. Liu, Linear recursion formulas of quantities of singular point and applications, Appl. Math. Comput., 2004, 148: 163–171.

[33] Y.R. Liu, J.B. Li, W.T. Huang, Singular point values, center problem and bifurcations of limit cycles of two dimensional differential autonomous systems, Nonlinear Dynamics Series, Vol. 6, Science Press, Beijing, 2008.

[34] Q. Wang, W. Huang, The equivalence between singular point quantities and Liapunov constants on center manifold, Adv. Diff. Equations, 2012, 2012: 78.

[35] B. Sang, Q. Wang, W. Huang, Computation of focal values and stability analysis of 4-dimensional systems, Electronic J. Diff. Equations, 2015, 2015(209): 1–11.

[36] F. Li, Y. Jin, Hopf bifurcation analysis and numerical simulation in a 4D-hyoerchaotic system, Nonlinear Dyn., 2012, 67(4): 2857–2864.

Section 2

Related Topics Pertaining to Manifolds

Chapter 2

An Intrinsic Characterization of Bonnet Surfaces Based on a Closed Differential Ideal

Paul Bracken

Additional information is available at the end of the chapter

http://dx.doi.org/10.5772/67008

Abstract

The structure equations for a two-dimensional manifold are introduced and two results based on the Codazzi equations pertinent to the study of isometric surfaces are obtained from them. Important theorems pertaining to isometric surfaces are stated and a theorem due to Bonnet is obtained. A transformation for the connection forms is developed. It is proved that the angle of deformation must be harmonic, and that the differentials of many of the important variables generate a closed differential ideal. This implies that a coordinate system exists in which many of the variables satisfy particular ordinary differential equations, and these results can be used to characterize Bonnet surfaces.

Keywords: manifold, differential form, closed, isometric, differential equation, Bonnet surface

1. Introduction

Bonnet surfaces in three-dimensional Euclidean space have been of great interest for a number of reasons as a type of surface [1, 2] for a long time. Bonnet surfaces are of nonconstant mean curvature that admits infinitely many nontrivial and geometrically distinct isometries, which preserve the mean curvature function. Nontrivial isometries are ones that do not extend to isometries of the whole space E^3. Considerable interest has resulted from the fact that the differential equations that describe the Gauss equations are classified by the type of related Painlevé equations they correspond to and they are integrated in terms of certain hypergeometric transcendents [3–5]. Here the approach first given by Chern [6] to Bonnet surfaces is considered. The development is accessible with many new proofs given. The main intention is to end by deriving an intrinsic characterization of these surfaces which indicates they are analytic. Moreover, it is shown that a type of Lax pair can be given for these surfaces and integrated. Several of the more important functions such as the mean curvature are seen to satisfy nontrivial ordinary differential equations.

Quite a lot is known about these surfaces. With many results the analysis is local and takes place under the assumptions that the surfaces contain no umbilic points and no critical points of the mean curvature function. The approach here allows the elimination of many assumptions and it is found the results are not too different from the known local ones. The statements and proofs have been given in great detail in order to help illustrate and display the interconnectedness of the ideas and results involved.

To establish some information about what is known, consider an oriented, connected, smooth open surface M in E^3 with nonconstant mean curvature function H. Moreover, suppose M admits infinitely many nontrivial and geometrically distinct isometries preserving H. Suppose U is the set of umbilic points of M and V the set of critical points of H. Many global facts are known with regard to U,V and H, and a few will now be mentioned. The set U consists of isolated points, even if there exists only one nontrivial isometry preserving the mean curvature, moreover, $U \subset V$ [7, 8]. Interestingly, there is no point in $V - U$ at which all order derivatives of H are zero, and V cannot contain any curve segment. If the function by which a nontrivial isometry preserving the mean curvature rotates the principal frame is considered, as when there are infinitely many isometries, this function is a global function on M continuously defined [9–11]. As first noted by Chern [6], this function is harmonic. The analysis will begin by formulating the structure equations for two-dimensional manifolds.

2. Structure equations

Over M, there exists a well-defined field of orthonormal frames, which is written as x, e_1, e_2, e_3 such that x$\in M$, e_3 is the unit normal at x, and e_1, e_2 are along principal directions [12]. The fundamental equations for M have the form

$$dx = \omega_1 e_1 + \omega_2 e_2, \quad de_1 = \omega_{12} e_2 + \omega_{13} e_3, \quad de_2 = -\omega_{12} e_1 + \omega_{23} e_3, \quad de_3 = -\omega_{13} e_1 - \omega_{23} e_2. \qquad (1)$$

Differentiating each of these equations in turn, results in a large system of equations for the exterior derivatives of the ω_i and ω_{ij}, as well as a final equation which relates some of the forms [13]. This choice of frame and Cartan's lemma allows for the introduction of the two principal curvatures which are denoted by a and c at x by writing

$$\omega_{12} = h\omega_1 + k\omega_2, \qquad \omega_{13} = a\omega_1, \qquad \omega_{23} = c\omega_2. \qquad (2)$$

Suppose that $a > c$ in the following. The mean curvature of M is denoted by H and the Gaussian curvature by K. They are related to a and c as follows

$$H = \frac{1}{2}(a + c), \qquad K = a \cdot c. \qquad (3)$$

The forms which appear in Eq. (1) satisfy the fundamental structure equations which are summarized here [14],

$$d\omega_1 = \omega_{12}\wedge\omega_2, \qquad d\omega_2 = \omega_1\wedge\omega_{12}$$

$$d\omega_{13} = \omega_{12}\wedge\omega_{23} \qquad d\omega_{23} = \omega_{13}\wedge\omega_{12}, \tag{4}$$

$$d\omega_{12} = ac\ \omega_2\wedge\omega_1 = -K\ \omega_1\wedge\omega_2.$$

The second pair of equations of (4) is referred to as the Codazzi equation and the last equation is the Gauss equation.

Exterior differentiation of the two Codazzi equations yields

$$(da-(a-c)h\omega_2)\wedge\omega_1 = 0, \qquad (dc-(a-c)k\omega_1)\wedge\omega_2 = 0. \tag{5}$$

Cartan's lemma can be applied to the equations in (5). Thus, there exist two functions u and v such that

$$\frac{1}{a-c}\ da-h\omega_2 = (u-k)\omega_1, \qquad \frac{1}{a-c}\ dc-k\omega_1 = (v-h)\omega_2. \tag{6}$$

Subtracting the pair of equations in (6) gives an expression for $d\log(a-c)$

$$d\log(a-c) = (u-2k)\ \omega_1-(v-2h)\ \omega_2. \tag{7}$$

Define the variable J to be

$$J = \frac{1}{2}(a-c) > 0. \tag{8}$$

It will appear frequently in what follows. Equation (7) then takes the form

$$d\log J = (u-2k)\omega_1-(v-2h)\omega_2. \tag{9}$$

The ω_i constitute a linearly independent set. Two related coframes called ϑ_i and α_i can be defined in terms of the ω_i and the functions u and v as follows,

$$\vartheta_1 = u\omega_1 + v\omega_2, \quad \vartheta_2 = -v\omega_1 + u\omega_2,$$
$$\alpha_1 = u\omega_1-v\omega_2, \quad \alpha_2 = v\omega_1 + u\omega_2. \tag{10}$$

These relations imply that $\vartheta_1 = 0$ is tangent to the level curves specified by H equals constant and $\alpha_1 = 0$ is its symmetry with respect to the principal directions.

Squaring both sides of the relation $2H = a + c$ and subtracting the relation $4K = 4ac$ yields $4(H^2-K) = (a-c)^2$. The Hodge operator, denoted by $*$, will play an important role throughout. It produces the following result on the basis forms ω_i,

$$*\omega_1 = \omega_2, \qquad *\omega_2 = -\omega_1, \qquad *^2 = -1. \tag{11}$$

Moreover, adding the expressions for da and dc given in Eq. (6), there results

$$\frac{1}{a-c}(da + dc) = (u-k)\omega_1 + h\omega_2 + (v-k)\omega_2 + k\omega_1 = u\omega_1 + v\omega_2 = \vartheta_1. \tag{12}$$

Finally, note that

$$\alpha_1 + 2 * \omega_{12} = u\omega_1 - v\omega_2 + 2 * (h\omega_1 + k\omega_2) = (u-2k)\omega_1 - (v-2h)\omega_2 = d\log J. \tag{13}$$

Therefore, the Codazzi equations (12) and (13) can be summarized using the definitions of H and J as

$$dH = J\vartheta_1, \qquad d\log J = \alpha_1 + 2 * \omega_{12}. \tag{14}$$

3. A theorem of Bonnet

Suppose that M^* is a surface which is isometric to M such that the principal curvatures are preserved [10–12]. Denote all quantities which pertain to M^* with the same symbols but with asterisks, as for example

$$a^* = a, \qquad c^* = c.$$

The same notation will be applied to the variables and forms which pertain to M and M^*. When M and M^* are isometric, the forms ω_i are related to the ω_i^* by the following transformation

$$\omega_1^* = \cos\tau\,\omega_1 - \sin\tau\,\omega_2, \qquad \omega_2^* = \sin\tau\,\omega_1 + \cos\tau\,\omega_2. \tag{15}$$

Theorem 3.1 Under the transformation of coframe given by Eq. (15), the associated connection forms are related by

$$\omega_{12}^* = \omega_{12} - d\tau. \tag{16}$$

Proof: Exterior differentiation of ω_1^* produces

$$d\omega_1^* = -\sin\tau\, d\tau\wedge\omega_1 + \cos\tau\, d\omega_1 - \cos\tau\, d\tau\wedge\omega_2 - \sin\tau\, d\omega_2$$
$$= d\tau\wedge(-\sin\tau\omega_1 - \cos\tau\omega_2) + \cos\tau\,\omega_{12}\wedge\omega_2 - \sin\tau\omega_1\wedge\omega_{12} = (-d\tau + \omega_{12})\wedge\omega_2^*.$$

Similarly, differentiating ω_2^* gives

$$d\omega_2^* = \cos\tau\wedge\omega_1 + \sin\tau\, d\omega_1 - \sin\tau\, d\tau\wedge\omega_2 + \cos\tau\, d\omega_2$$
$$= d\tau\wedge(\cos\tau\omega_1 - \sin\tau\omega_2) + \sin\tau\omega_{12}\wedge\omega_2 + \cos\tau\omega_1\wedge\omega_{12} = \omega_1^*\wedge(-d\tau + \omega_{12}).$$

There is a very important result that can be developed at this point. In the case that $a = a^*$ and $c = c^*$, the Codazzi equations imply that

$$\alpha_1 + 2 * \omega_{12} = d\log(a - c) = d\log(a^* - c^*) = \alpha_1^* + 2 * \ \omega_{12}^*.$$

Apply the operator $*$ to both sides of this equation, we obtain

$$\alpha_2 - 2\omega_{12} = \alpha_2^* - 2\omega_{12}^*.$$

Substituting for ω_{12}^* from Theorem 3.1, this is

$$2d\tau = \alpha_2 - \alpha_2^*. \tag{17}$$

Lemma 3.1

$$\vartheta_1 = \vartheta_1^*.$$

Proof: This can be shown in two ways. First from Eq. (15), express the ω_i in terms of the ω_i^*

$$\omega_1 = \cos\tau\,\omega_1^* + \sin\tau\,\omega_2^*, \qquad \omega_2 = -\sin\tau\,\omega_1^* + \cos\tau\,\omega_2^*. \tag{18}$$

Therefore,

$$\vartheta_1 = u\omega_1 + v\omega_2 = u(\cos\tau\,\omega_1^* + \sin\tau\omega_2^*) + v(-\sin\tau\omega_1^* + \cos\tau\omega_2^*) = u^*\omega_1^* + v^*\omega_2^* = \vartheta_1^*,$$

where $u^* = u\cos\tau - v\sin\tau$ and $v^* = u\sin\tau + v\cos\tau$. □

Lemma 3.1 also follows from the fact that $dH = dH^*$ and Eq. (8).

Lemma 3.2

$$\alpha_2^* = \sin(2\tau)\,\alpha_1 + \cos(2\tau)\,\alpha_2.$$

Proof:

$$\begin{aligned}\alpha_2^* &= (u\sin\tau + v\cos\tau)(\cos\tau\omega_1 - \sin\tau\omega_2) + (u\cos\tau - v\sin\tau)(\sin\tau\omega_1 + \cos\tau\omega_2)\\ &= (u\sin(2\tau) + v\cos(2\tau))\omega_1 + (-v\sin(2\tau) + u\cos(2\tau))\omega_2\\ &= \sin(2\tau)\alpha_1 + \cos(2\tau)\alpha_2.\end{aligned}$$

Substituting α_2^* from Lemma 3.2 into Eq. (13), $d\tau$ can be written as

$$d\tau = \frac{1}{2}(\alpha_2 - \sin(2\tau)\alpha_1 - \cos(2\tau)\alpha_2) = \frac{1}{2}((1 - \cos(2\tau))\alpha_2 - \sin(2\tau)\alpha_1). \tag{19}$$

Introduce the new variable $t = \cot(\tau)$ so $dt = -\csc^2(\tau)\,d\tau$ and $\sin\tau = \frac{1}{\sqrt{1+t^2}}$, $\cos\tau = \frac{1}{\sqrt{1+t^2}}$, hence the following lemma.

Lemma 3.3

$$dt = t\alpha_1 - \alpha_2.$$

This is the total differential equation which must be satisfied by the angle τ of rotation of the principal directions during the deformation. If the deformation is to be nontrivial, it must be that this equation is completely integrable.

Theorem 3.2 A surface M admits a nontrivial isometric deformation that keeps the principal curvatures fixed if and only if

$$d\alpha_1 = 0, \qquad d\alpha_2 = \alpha_1 \wedge \alpha_2. \tag{20}$$

Proof: Differentiating both sides of Lemma 3.3 gives

$$dt \wedge \alpha_1 + td\alpha_1 - d\alpha_2 = (t\alpha_1 - \alpha_2)\wedge\alpha_1 + td\alpha_1 - d\alpha_2 = 0.$$

Equating the coefficients of t to zero gives the result (20).

This theorem seems to originate with Chern [6] and is very useful because it gives the exterior derivatives of the α_i. When the mean curvature is constant, $dH = 0$, hence it follows from Eq. (14) that $\vartheta_1 = 0$. This implies that $u = v = 0$, and so α_1 and α_2 must vanish. Hence, $dt = 0$ which implies that, since the α_i is linearly independent, t equals a constant. Thus, we arrive at a theorem originally due to Bonnet.

Theorem 3.3 A surface of constant mean curvature can be isometrically deformed preserving the principal curvatures. During the deformation, the principal directions rotate by a fixed angle.

4. Connection form associated to a coframe and transformation properties

Given the linearly independent one forms ω_1,ω_2, the first two of the structure equations uniquely determine the form ω_{12}. The ω_1,ω_2 is called the orthonormal coframe of the metric

$$ds^2 = \omega_1^2 + \omega_2^2,$$

and ω_{12} is the connection form associated with it.

Theorem 4.1 Suppose that $A > 0$ is a function on M. Under the change of coframe

$$\omega_1^* = A\omega_1, \qquad \omega_2^* = A\omega_2, \tag{21}$$

the associated connection forms are related by

$$\omega_{12}^* = \omega_{12} + * \, d\log A. \tag{22}$$

Proof: The structure equations for the transformed system are given as

$$d\omega_1^* = \omega_{12}^* \wedge \omega_2^*, \qquad d\omega_2^* = \omega_1^* \wedge \omega_{12}^*.$$

Using Eq. (21) to replace the ω_i^* in these, we obtain

$$d\log A \wedge \omega_1 + d\omega_1 = \omega_{12}^* \wedge \omega_2, \qquad d\log A \wedge \omega_2 + d\omega_2 = \omega_1 \wedge \omega_{12}^*.$$

The ω_i satisfy a similar system of structure equations, so replacing $d\omega_i$ here yields

$$(\omega_{12}^* - \omega_{12}) \wedge \omega_2 = d\log A \wedge \omega_1, \qquad (\omega_{12}^* - \omega_{12}) \wedge \omega_1 = -d\log A \wedge \omega_2.$$

Since the form ω_i satisfies the equations $*\omega_1 = \omega_2$ and $*\omega_2 = -\omega_1$, substituting these relations into the above equations and using $\Omega_k \wedge (*\Theta_k) = \Theta_k \wedge (*\Omega_k)$, we obtain that in the form

$$\omega_1 \wedge *(\omega_{12}^* - \omega_{12}) = -\omega_1 \wedge d\log A, \qquad \omega_2 \wedge *(\omega_{12}^* - \omega_{12}) = -\omega_2 \wedge d\log A.$$

Cartan's lemma can be used to conclude from these that there exist functions f and g such that

$$*(\omega_{12}^* - \omega_{12}) = -d\log A - f\omega_1, \qquad *(\omega_{12}^* - \omega_{12}) = -d\log A + g\omega_2.$$

Finally, apply $*$ to both sides and use $*^2 = -1$ to obtain

$$\omega_{12}^* - \omega_{12} = *d\log A + f\omega_2, \qquad \omega_{12}^* - \omega_{12} = *d\log A + g\omega_1.$$

The forms ω_i are linearly independent, so for these two equations to be compatible, it suffices to put $f = g = 0$, and the result follows. □

For the necessity in the Chern criterion, Theorem 3.2, no mention of the set V of critical points of H is needed. In fact, when H is constant, this criterion is met and the sufficiency also holds with τ constant. However, when H is not identically constant, we need to take the set V of critical points into account for the sufficiency. In this case, $M - V$ is also an open, dense, and connected subset of M. On this subset $J > 0$ and the function A can be defined in terms of the functions u and v as

$$A = +\sqrt{u^2 + v^2} > 0. \tag{23}$$

To define more general transformations of the ω_i, define the angle ψ as

$$u = A\cos(\psi), \qquad v = A\sin(\psi). \tag{24}$$

This angle, which is defined modulo 2π, is continuous only locally and could be discontinuous in a nonsimply connected region of $M - V$. With A and ψ related to u and v by Eq. (24), the forms ϑ_i and α_i can be written in terms of A and ψ as

$$\vartheta_1 = A(\cos(\psi)\,\omega_1 + \sin(\psi)\omega_2), \quad \vartheta_2 = A(-\sin(\psi)\,\omega_1 + \cos(\psi)\,\omega_2),$$
$$\alpha_1 = A(\cos(\psi)\,\omega_1 - \sin(\psi)\,\omega_2), \quad \alpha_2 = A(\sin(\psi)\,\omega_1 + \cos(\psi)\,\omega_2). \tag{25}$$

The forms ω_i, ϑ_i, α_i define the same structure on M and we let ω_{12}, ϑ_{12}, α_{12} be the connection forms associated to the coframes ω_1,ω_2; ϑ_1,ϑ_2; α_1,α_2. The next theorem is crucial for what follows.

Theorem 4.2

$$\vartheta_{12} = d\psi + \omega_{12} + *d\log A = 2d\psi + \alpha_{12}. \tag{26}$$

Proof: Each of the transformations which yield the ϑ_i and α_i in the form (25) can be thought of as a composition of the two transformations which occur in the Theorems 3.1 and 4.1. First apply the transformation $\omega_i \to A\omega_i$ and $\tau \to -\psi$ with $\omega_i^* \to \vartheta_i$ in Eq. (15), we get the ϑ_i equations in Eq. (25). Invoking Theorems 3.1 and 4.1 in turn, the first result is obtained

$$\vartheta_{12} = d\psi + \omega_{12} + *\,d\log A.$$

The transformation to the α_i is exactly similar except that $\tau \to \psi$, hence

$$\alpha_{12} = -d\psi + \omega_{12} + \;*\; d\log A.$$

This implies $*d\log A = \alpha_{12} + d\psi - \omega_{12}$. When replaced in the first equation of (26), the second equation appears. Note that from Theorem 3.2, $\alpha_{12} = \alpha_2$, so the second equation can be given as $\vartheta_{12} = 2d\psi + \alpha_2$.

Differentiating the second equation in Eq. (14) and using $d\alpha_1 = 0$, it follows that

$$d\,*\,\omega_{12} = 0. \tag{27}$$

Lemma 4.1 The angle ψ is a harmonic function $d * d\psi = 0$ and moreover, $d * \vartheta_{12} = 0$.

Proof: From Theorem 4.2, it follows by applying $*$ through Eq. (26) that

$$*\vartheta_{12} = *\omega_{12} + *d\psi - d\log A = 2 * d\psi - \alpha_1. \tag{28}$$

Exterior differentiation of this equation using $d * \omega_{12} = 0$ immediately gives

$$d * d\psi = 0.$$

This states that ψ is a harmonic function. Equation (28) also implies that $d * \vartheta_{12} = 0$.

5. Construction of the closed differential ideal associated with M

Exterior differentiation of the first equation in (14) and using the second equation produces

$$d\vartheta_1 + (\alpha_1 + 2 * \omega_{12})\wedge\vartheta_1 = 0. \tag{29}$$

The structure equation for the ϑ_i will be needed,

$$d\vartheta_1 = \vartheta_{12}\wedge\vartheta_2 = - * \vartheta_{12}\wedge\vartheta_1. \tag{30}$$

From the second equation in Eq. (26), we have $*\omega_{12}-d\log A + \alpha_1 = *d\psi$, and putting this in the first equation of Eq. (26), we find

$$- * \vartheta_{12} + \alpha_1 + 2 \; * \omega_{12} = 2\, d \log A. \tag{31}$$

Using Eq. (31) in Eq. (30),

$$d\vartheta_1 + (\alpha_1 + 2 * \omega_{12})\wedge\vartheta_1 = 2\, d \log A\wedge\vartheta_1. \tag{32}$$

Replacing $d\vartheta_1$ by means of Eq. (29) implies the following important result

$$d \log A\wedge\vartheta_1 = 0. \tag{33}$$

Equation (33) and Cartan's lemma imply that there exists a function B such that

$$d \log A = B\vartheta_1. \tag{34}$$

This is the first in a series of results which relates many of the variables in question such as J, B, and ϑ_{12} directly to the one-form ϑ_1. To show this requires considerable work. The way to proceed is to use the forms α_i in Theorem 3.2 because their exterior derivatives are known. For an arbitrary function on M, define

$$df = f_1\alpha_1 + f_2\alpha_2. \tag{35}$$

Differentiating Eq. (35) and extracting the coefficient of $\alpha_1\wedge\alpha_2$, we obtain

$$f_{21} - f_{12} + f_2 = 0. \tag{36}$$

In terms of the α_i, $*d\psi = \psi_1\alpha_2 - \psi_2\alpha_1$, Lemma 4.1 yields

$$\psi_{11} + \psi_{22} + \psi_1 = 0. \tag{37}$$

Finally, since $*\vartheta_{12} = 2 * d\psi - \alpha_1$, substituting for $*d\psi$, we obtain that

$$*\vartheta_{12} = -(2\psi_2 + 1)\alpha_1 + 2\psi_1\,\alpha_2. \tag{38}$$

Differentiating structure equation (30) and using Lemma 4.1,

$$*\vartheta_{12}\wedge\, d\vartheta_1 = 0,$$

so,

$$*\vartheta_{12}\wedge\vartheta_{12}\wedge\vartheta_2 = 0$$

This equation implies that either ϑ_{12} or $*\vartheta_{12}$ is a multiple by a function of the form ϑ_2. Hence, for some function p,

$$\begin{aligned} \vartheta_{12} = -p\vartheta_2, \quad *\vartheta_{12} = p\vartheta_1, \\ \vartheta_{12} = p\vartheta_1, \quad *\vartheta_{12} = p\vartheta_2, \end{aligned} \tag{39}$$

Substituting the first line of Eq. (39) back into the structure equation, we have

$$d\vartheta_1 = 0. \tag{40}$$

The second line yields simply $d\vartheta_1 = p\vartheta_1\wedge\vartheta_2$. Only the first case is examined now. Substituting Eq. (40) into Eq. (29), the following important constraint is obtained

$$(\alpha_1 + 2 * \omega_{12})\wedge\vartheta_1 = 0. \tag{41}$$

Theorem 5.1 The function ψ satisfies the equation

$$2\psi_1\cos(2\psi) + (2\psi_2 + 1)\sin(2\psi) = 0. \tag{42}$$

Proof: By substituting $*d\psi$ into Eq. (28) we have

$$*\vartheta_{12} = 2 * (\psi_1\alpha_1 + \psi_2\alpha_2) - \alpha_1 = -(2\psi_2 + 1)\alpha_1 + 2\psi_1\alpha_2. \tag{43}$$

Substituting Eq. (43) into Eq. (26) and solving for $*\omega_{12}$, we obtain that

$$*\omega_{12} = *\vartheta_{12} - * d\psi + d\log A = *\vartheta_{12} - * d\psi + B\vartheta_1 = *d\psi - \alpha_1 + B\vartheta_1.$$

This can be put in the equivalent form

$$2 * \omega_{12} + \alpha_1 = 2 * d\psi - \alpha_1 + 2B\vartheta_1. \tag{44}$$

Taking the exterior product with ϑ_1 and using $d\psi_1$, we get

$$(\alpha_1 + 2*\omega_{12})\wedge\vartheta_1 = (2*d\psi-\alpha_1)\wedge\vartheta_1 = (2\psi_1*\alpha_1 + 2\psi_2*\alpha_2-\alpha_1)\wedge\vartheta_1$$
$$= (2\psi_1\cos(2\psi) + (2\psi_2+1)\sin(2\psi))\vartheta_2\wedge\vartheta_1.$$

Imposing the constraint (41), the coefficient of $\vartheta_1\wedge\vartheta_2$ can be equated to zero. This produces the result (42).

As a consequence of Theorem 5.1, a new function C can be introduced such that

$$2\psi_1 = C\sin(2\psi), \qquad 2\psi_2 + 1 = -C\cos(2\psi). \tag{45}$$

Differentiation of each of these with respect to the α_i basis, we get for $i = 1, 2$ that

$$2\psi_{1i} = C_i\sin(2\psi) + 2\psi_i\,C\cos(2\psi), \qquad 2\psi_{2i} = -C_i\cos(2\psi) + 2\psi_i\,C\sin(2\psi).$$

Substituting $f = \psi$ into Eq. (36) and using the fact that ψ satisfies Eq. (37) gives the pair of equations

$$-C_1\cos(2\psi)-C_2\sin(2\psi) + 2\psi_1 C\sin(2\psi)-(2\psi_2+1)C\cos(2\psi)-1 = 0,$$
$$C_1\sin(2\psi)-C_2\cos(2\psi) + 2\psi_1 C\cos(2\psi) + (2\psi_2+1)C\sin(2\psi) = 0.$$

This linear system can be solved for C_1 and C_2 to get

$$C_1 + C(2\psi_2+1) + \cos(2\psi) = 0, \quad C_2-2C\psi_1 + \sin(2\psi) = 0. \tag{46}$$

By differentiating each of the equations in (46), it is easy to verify that C satisfies Eq. (36), namely, $C_{12}-C_{21}-C_2 = 0$. Hence, there exist harmonic functions which satisfy Eq. (42). The solution depends on two arbitrary constants, the values of ψ and C at an initial point.

Lemma 5.1

$$dC = (C^2-1)\vartheta_1, \qquad *\vartheta_{12} = C\vartheta_1. \tag{47}$$

Proof: It is easy to express the ϑ_i in terms of the α_i,

$$\vartheta_1 = \cos(2\psi)\alpha_1 + \sin(2\psi)\alpha_2, \qquad \vartheta_2 = -\sin(2\psi)\alpha_1 + \cos(2\psi)\alpha_2. \tag{48}$$

Therefore, using Eqs. (45) and (46), it is easy to see that

$$dC = C_1\alpha_1 + C_2\alpha_2 = (C^2-1)(\cos(2\psi)\alpha_1 + \sin(2\psi)\alpha_2) = (C^2-1)\vartheta_1.$$

Using Eq. (45), it follows that

$$*\vartheta_{12} = -(2\psi_2+1)\alpha_1 + 2\psi_1\alpha_2 = C\cos(2\psi)\alpha_1 + C\sin(2\psi)\alpha_2$$
$$= C(\cos(2\psi)\alpha_1 + \sin(2\psi)\alpha_2) = C\vartheta_1.$$

This implies that $\vartheta_{12} = -C\vartheta_2$.

It is possible to obtain formulas for B_1, B_2. Using Eq. (48) in Eq. (34), the derivatives of $\log A$ can be written down

$$(\log A)_1 = B\cos(2\psi), \qquad (\log A)_2 = B\sin(2\psi). \tag{49}$$

Differentiating each of these in turn, we obtain for $i = 1, 2$,

$$(\log A)_{1i} = B_i\cos(2\psi) - 2B\psi_i\sin(2\psi), \qquad (\log A)_{2i} = B_i\sin(2\psi) + 2B\psi_i\cos(2\psi). \tag{50}$$

Taking $f = \log A$ in Eq. (36) produces a first equation for the B_i,

$$B_1\sin(2\psi) + 2B\psi_1\cos(2\psi) - B_2\cos(2\psi) + 2B\psi_2\sin(2\psi) + B\sin(2\psi) = 0. \tag{51}$$

If another equation in terms of B_1 and B_2 can be found, it can be solved simultaneously with Eq. (51). There exists such an equation and it can be obtained from the Gauss equation in (4) which we put in the form

$$d\omega_{12} = -ac\ \omega_1 \wedge \omega_2 = -ac\ A^{-2}\ \alpha_1 \wedge \alpha_2.$$

Solving Eq. (26) for ω_{12}, we have

$$\omega_{12} = d\psi + \alpha_2 + (\log A)_2\alpha_1 - (\log A)_1\alpha_2.$$

The exterior derivative of this takes the form,

$$d\omega_{12} = [1 - (\log A)_{11} - (\log A)_{22} - (\log A)_1]\alpha_1 \wedge \alpha_2.$$

Putting this in the Gauss equation,

$$-(\log A)_{11} - (\log A)_{22} + \{-(\log A)_1 + 1\} + acA^{-2} = 0.$$

Replacing the second derivatives from Eq. (50), we have the required second equation

$$-B_1\cos(2\psi) - B_2\sin(2\psi) + B\{2\psi_1\sin(2\psi) - (2\psi_2 + 1)\cos(2\psi)\} + 1 + acA^{-2} = 0. \tag{52}$$

Solving Eqs. (51) and (52) together, the following expressions for B_1 and B_2 are obtained

$$B_1 + B(2\psi_2 + 1) - (1 + acA^{-2})\cos(2\psi) = 0, \qquad B_2 - 2B\psi_1 - (1 + acA^{-2})\sin(2\psi) = 0. \tag{53}$$

Given these results for B_1 and B_2, it is easy to produce the following two Lemmas.

Lemma 5.2

$$dB = (BC + 1 + acA^{-2})\vartheta_1, \qquad d\log J = (C + 2B)\vartheta_1. \tag{54}$$

Proof: Substituting Eq. (53) into dB, we get

$$dB = B_1\alpha_1 + B_2\alpha_2 = (BC + 1 + acA^{-2})(\cos(2\psi)\alpha_1 + \sin(2\psi)\alpha_2) = (BC + 1 + acA^{-2})\,\vartheta_1.$$

Moreover,

$$d\log J = \alpha_1 + 2*\omega_{12} = \alpha_1 + 2(*\vartheta_{12} - *d\psi + d\log A) = \alpha_1 + 2*\vartheta_{12} - 2*d\psi + 2d\log A$$
$$= *\vartheta_{12} + 2d\log A = C\vartheta_1 + 2B\vartheta_1.$$

Lemma 5.3

$$d\psi = -\frac{1}{2}\sin(2\psi)\vartheta_1 - \frac{1}{2}(C + \cos(2\psi))\vartheta_2. \tag{55}$$

Proof:

$$2d\psi = 2\psi_1\alpha_1 + 2\psi_2\alpha_2 = C\sin(2\psi)\alpha_1 - (C\cos(2\psi) + 1)\alpha_2$$
$$= C\sin(2\psi)(\cos(2\psi)\vartheta_1 - \sin(2\psi)\vartheta_2) - (C\cos(2\psi) + 1)(\sin(2\psi)\vartheta_1 + \cos(2\psi)\vartheta_2)$$
$$= -\sin(2\psi)\vartheta_1 - (C + \cos(2\psi))\vartheta_2.$$

In the interests of completeness, it is important to verify the following theorem.

Theorem 5.2 The function B satisfies Eq. (36) provided ψ satisfies both Eqs. (37) and (41).

Proof: Differentiating B_1 and B_2 given by Eq. (53), the left side of Eq. (36) is found to be

$$B_{21} - B_{12} + B_2 = 2B_1\psi_1 + B_2(2\psi_2 + 1) + 2B(\psi_{11} + \psi_{22} + \psi_1) + A^{-2}((ac)_1\sin(2\psi) - (ac)_2\sin(2\psi))$$
$$-2acBA^{-2}(\cos(2\psi)\sin(2\psi) - \sin(2\psi)\cos(2\psi)) + (1 + acA^{-2})(2\psi_1\cos(2\psi) + (2\psi_2 + 1)\sin(2\psi))$$
$$= 2(1 + acA^{-2})(2\psi_1\cos(2\psi) + (2\psi_2 + 1)\sin(2\psi)) + A^{-2}((ac)_1\sin(2\psi) - (ac)_2\cos(2\psi)).$$

To simplify this, Eq. (37) has been substituted. Using Eq. (48) and $*d(ac) = (ac)_1\alpha_2 - (ac)_2\alpha_1$, it follows that

$$*d(ac)\wedge\vartheta_2 = ((ac)_1\sin(2\psi) - (ac)_2\cos(2\psi))\alpha_1\wedge\alpha_2.$$

Note that the coefficient of $\alpha_1\wedge\alpha_2$ in this appears in the compatibility condition. To express it in another way, begin by finding the exterior derivative of $4ac = (a + c)^2 - (a - c)^2$,

$$4d(ac) = 2(a + c)(a - c)\vartheta_1 - 2(a - c)^2(\alpha_1 + 2*\omega_{12}).$$

Applying the Hodge operator to both sides of this, gives upon rearranging terms

$$2*\frac{d(ac)}{a - c} = (a + c)\vartheta_2 - (a - c)(\alpha_2 - 2\omega_{12}).$$

Consequently, we can write

$$-\frac{2}{(a - c)^2}*d(ac)\wedge\vartheta_2 = (\alpha_2 - 2\omega_{12})\wedge\vartheta_2 = -(2\psi_1\cos(2\psi) + (2\psi_2 + 1)\sin(2\psi))\alpha_1\wedge\alpha_2.$$

Therefore, it must be that

$$-(ac)_1 \sin(2\psi) + (ac)_2 \cos(2\psi) = -\frac{1}{2}(a-c)^2(2\psi_1 \cos(2\psi) + (2\psi_2 + 1)\sin(2\psi)).$$

It follows that when $f = B$, Eq. (36) finally reduces to the form

$$(1 + H^2 A^{-2})[2\psi_1 \cos(2\psi) + (2\psi_2 + 1)\sin(2\psi)] = 0.$$

The first factor is clearly nonzero, so the second factor must vanish. This of course is equivalent to the constraint (41).

6. Intrinsic characterization of *M*

During the prolongation of the exterior differential system, the additional variables ψ, A, B, and C have been introduced. The significance of the appearance of the function C, is that the process terminates and the differentials of all these functions can be computed without the need to introduce more functions. This means that the exterior differential system has finally closed.

The results of the previous section, in particular, the lemmas, can be collected such that they justify the following.

Proposition 6.1 The differential system generated in terms of the differentials of the variables ψ, A, B, and C is closed. The variables H, J, A, B, C remain constant along the ϑ_2-curves so $\vartheta_1 = 0$. Hence, an isometry that preserves H must map the ϑ_1, ϑ_2 curves onto the corresponding ϑ_1^*, ϑ_2^* curves of the associated surface M^* which is isometric to M.

Along the ϑ_1, ϑ_2 curves, consider the normalized frame,

$$\zeta_1 = \cos(\psi)e_1 + \sin(\psi)e_2, \qquad \zeta_2 = -\sin(\psi)e_1 + \cos(\psi)e_2. \tag{56}$$

The corresponding coframe and connection form are

$$\xi_1 = \cos(\psi)\omega_1 + \sin(\psi)\omega_2, \quad \xi_2 = -\sin(\psi)\omega_1 + \cos(\psi)\omega_2, \quad \xi_{12} = d\psi + \omega_{12}. \tag{57}$$

Then ϑ_1 can be expressed as a multiple of ξ_1 and ϑ_2,ϑ_{12} in terms of ξ_2, and the differential system can be summarized here:

$$\begin{gathered}\vartheta_1 = A\xi_1, \qquad \vartheta_2 = A\xi_2, \quad \vartheta_{12} = \xi_{12} + *d\log A = -CA\xi_2,\\ d\log A = AB\xi_1, \qquad dB = A(BC + 1 + acA^{-2})\xi_1, \qquad dC = A(C^2-1)\xi_1,\\ dH = AJ\xi_1, \quad dJ = AJ(2B + C)\xi_1.\end{gathered} \tag{58}$$

The condition $d\vartheta_1 = 0$ is equivalent to

$$dA \wedge \xi_1 + Ad\xi_1 = 0.$$

This implies that $d\xi_1 = 0$ since dA is proportional to ξ_1. Also, $d * \vartheta_{12} = 0$ is equivalent to $d * \xi_{12} = 0$.

Moreover, $d * \xi_{12} = 0$ is equivalent to the fact that the ξ_1,ξ_2 curves can be regarded as coordinate curves parameterized by isothermal parameters. Therefore, along the ξ_1,ξ_2 curves, orthogonal isothermal coordinates denoted (s,t) can be introduced. The first fundamental form of M then takes the form,

$$I = \xi_1^2 + \xi_2^2 = E(s)(ds^2 + dt^2). \tag{59}$$

Now suppose we set $e(s) = \sqrt{E(s)}$, then

$$\xi_1 = e(s)\, ds, \qquad \xi_2 = e(s)\, dt, \qquad \xi_{12} = \frac{e'(s)}{e^2(s)}\xi_2 = \frac{e'(s)}{e(s)}\, dt. \tag{60}$$

This means such a surface is isometric to a surface of revolution. Since ψ, $d^*\xi_{12} = 0$, Eq. (57) implies that $d^*\omega_{12} = 0$. This can be stated otherwise as the principal coordinates are isothermal and so M is an isothermic surface.

Since A, B, C, H, and J are functions of only the variable s, this implies that H and J, or H and K, are constant along the t curves where s is constant. This leads to the following proposition.

Proposition 6.2

$$dH \wedge dK = 0, \qquad \xi_{12} = -(C + B)A\xi_2. \tag{61}$$

This is equivalent to the statement M is a Weingarten surface.

Proof: The first result follows from the statement about the coordinate system above. Since $\vartheta_{12} = \xi_{12} + *d\log A = -CA\xi_2$ and $dA = A^2 B\xi_1$,

$$\xi_{12} = -CA\xi_2 - * \, d\log A = -CA\xi_2 - * \, A^{-1}\, dA = -CA\xi_2 - AB * \xi_1 = -(C + B)A\xi_2$$

Consequently, the geodesic curvature of each ξ_2 curve, s constant, is

$$\frac{e'(s)}{e^2(s)} = -A(B + C),$$

which is constant.

To express the ω_i in terms of ds and dt, start by writing ω_i in terms of the ξ_i and then substituting Eq. (60),

$$\omega_1 = \cos(\psi)e\, ds - \sin(\psi)e\, dt, \qquad \omega_2 = \sin(\psi)e\, ds + \cos(\psi)e\, dt. \tag{62}$$

Subscripts (s,t) denote differentiation and $H_s = H'$ is used interchangeably. Beginning with $dH = H'\, ds$ and using Eq. (62), we have

$$dH = H_1\omega_1 + H_2\omega_2 = (H_1 \cos(\psi) + H_2 \sin(\psi))\, e\, ds + (-H_1 \sin(\psi) + H_2 \cos(\psi))\, e\, dt = H'\, ds.$$

Equating coefficients of differentials, this implies that

$$H_1 e \cos(\psi) + H_2 e \sin(\psi) = H', \qquad -H_1 \sin(\psi) + H_2 \cos(\psi) = 0.$$

Solving this as a linear system we obtain H_1, H_2,

$$H_1 = \frac{H'}{e} \cos(\psi), \qquad H_2 = \frac{H'}{e} \sin(\psi). \tag{63}$$

Noting that $u = H_1/J$ and $v = H_2/J$, using Eq. (57) the forms α_i can be expressed in terms of ds, dt

$$\alpha_1 = \frac{H'}{J}(\cos(2\psi)\, ds - \sin(2\psi)\, dt), \qquad \alpha_2 = \frac{H'}{J}(\sin(2\psi)\, ds + \cos(2\psi)\, dt). \tag{64}$$

Substituting ξ_1 from Eq. (60) into $dH = AJ\xi_1$,

$$dH = H' ds = AJ\xi_1 = AJ\, e(s)\, ds.$$

Therefore, $H' = AJe > 0$ and so $H(s)$ is an increasing function of s. Now define the function $Q(s)$ to be

$$Q = \frac{H'}{J} = A \cdot e > 0. \tag{65}$$

Substituting Eq. (65) into Eq. (64), α_i is expressed in terms of Q as well. Equations (20) in Theorem 3.2 can easily be expressed in terms of ψ and Q.

Theorem 6.1 Equation (20) is equivalent to the following system of coupled equations in ψ and Q:

$$\begin{aligned} \sin(2\psi)(\log(Q))_s + 2\cos(2\psi)\psi_s - 2\sin(2\psi)\psi_t &= 0, \\ \cos(2\psi)(\log(Q))_s - 2\sin(2\psi)\psi_s - 2\cos(2\psi)\psi_t &= Q. \end{aligned} \tag{66}$$

Moreover, Eq. (66) is equivalent to the following first-order system

$$\psi_s = -\frac{1}{2} Q \sin(2\psi), \qquad \psi_t = \frac{1}{2}(\log(Q))_s - \frac{1}{2} Q \cos(2\psi). \tag{67}$$

System (67) can be thought of as a type of Lax pair. Moreover, Eq. (67) implies that ψ is harmonic as well. Differentiating ψ_s with respect to s and ψ_t with respect to t, it is clear that ψ satisfies Laplace's equation in the (s,t) variables $\psi_{ss} + \psi_{tt} = 0$. This is another proof that ψ is harmonic.

Theorem 6.2 The function $Q(s)$ satisfies the following second-order nonlinear differential equation

$$Q''(s)Q(s)-(Q'(s))^2 = Q^4(s). \quad (68)$$

There exists a first integral for this equation of the following form

$$Q'(s)^2 = Q(s)^4 + \kappa Q(s)^2, \quad \kappa \in \mathbb{R}. \quad (69)$$

Proof: Equation (68) is just the compatibility condition for the first-order system (67). The required derivatives are

$$\psi_{st} = -\frac{Q}{2}\cos(2\psi)((\log Q)_s - Q\cos(2\psi)), \quad \psi_{ts} = \frac{1}{2}(\log Q)_{ss} - \frac{1}{2}Q_s\cos(2\psi) + Q\sin(2\psi)\psi_s.$$

Equating derivatives $\psi_{st} = \psi_{ts}$, the required (68) follows.

Differentiating both sides of Eq. (69) we get

$$Q''(s) = 2Q(s)^3 + \kappa Q(s). \quad (70)$$

Isolating $\kappa Q(s)$ from Eq. (69) and substituting it into Eq. (70), Eq. (68) appears.

It is important to note that the function C which appears when the differential ideal closes can be related to the function Q.

Corollary 6.1

$$C = \left(\frac{1}{Q}\right)'. \quad (71)$$

Proof: Using ϑ_i from Eq. (58) in Lemma 5.3, in the s,t coordinates

$$2d\psi = -\sin(2\psi)\, Ae\, ds - (C + \cos(2\psi))\, Ae\, dt = \psi_s ds + \psi_t\, dt$$

Hence using Eq. (67), this implies that $2\psi_s = -\sin(2\psi)\, Ae = -Q\sin(2\psi)$, hence $Q = Ae$. The second equation in Eq. (67) for ψ_t implies that $(C + \cos(2\psi))\, Ae = Q\cos(2\psi) - (\log Q)'$. Replacing $Ae = Q$, this simplifies to the form (71).

7. Integrating the Lax pair system

It is clear that the first-order equation in (67) for $Q(s)$ is separable and can be integrated. The integral depends on whether K is zero or nonzero:

$$Q(s) = \frac{1}{\varepsilon s + \gamma}, \quad K = 0; \qquad \log\left(\frac{2(K + \sqrt{K}\sqrt{Q^2 + K})}{Q}\right) = \varepsilon\sqrt{K}s + \gamma, \quad K \neq 0. \quad (72)$$

Here $\varepsilon = \pm 1$ and γ is the last constant of integration. Taking specific choices for the constants, for example, $e^{\gamma} = 2\sqrt{K}$ when $K \neq 0$ and $a = \sqrt{K}$, the set of solutions (72) for $Q(s)$ can be summarized below.

$Dom(s)$	$Q(s)$	$Dom(s)$	$Q(s)$
$s > 0$	$\frac{1}{s}$	$s < 0$	$-\frac{1}{s}$
$0 < s < \frac{\pi}{a}$	$\frac{a}{\sin(as)}$	$-\frac{\pi}{a} < s < 0$	$-\frac{a}{\sin(as)}$
$s > 0$	$\frac{a}{\sinh(as)}$	$s < 0$	$-\frac{a}{\sinh(as)}$

(73)

It is presumed that other choices of the constants can be geometrically eliminated in favor of Eq. (73). The solutions (73) are then substituted back into linear system (67). The first equation in (67) implies that either

$$\psi \equiv 0, \quad \mathrm{mod}\frac{\pi}{2}; \qquad \frac{2\psi_s}{\sin(2\psi)} = -Q. \tag{74}$$

Substitute $\psi \equiv 0$ into the second equation in (67). It implies that $(\log Q)_s = Q$ and $\psi = \pi/2$ gives $(\log Q)_s = -Q$. In both cases $Q(s)$ is a solution which already appears in Eq. (73).

For the second case in Eq. (74), the equation can be put in the form

$$(\log|\tan(\psi)|)_s = -Q.$$

Integrating we have for some function $y(t)$ to be determined,

$$\tan(\psi) = e^{-\int Q(s)ds} \cdot y(t). \tag{75}$$

Therefore, $\tan(\psi)$ can be obtained by substituting for $Q(s)$ for each of the three cases in Eq. (73). The upper sign holds for $s > 0$ and the lower sign holds if $s < 0$.

i. $Q(s) = \pm s^{-1}, -\int Q(s)ds = \log|s|^{\mp}$ and

$$\tan(\psi) = s^{\mp} \cdot y(t). \tag{76}$$

ii. $Q(s) = \pm\frac{a}{\sin(as)}, -\int Q(s)ds = \log|\csc(as) - \cot(as)|^{\mp}$ and

$$\tan(\psi) = \left(\tan\left(\frac{as}{2}\right)\right)^{\mp} \cdot y(t). \tag{77}$$

iii. $Q(s) = \pm\frac{a}{\sinh(as)}, -\int Q(s)ds = \mp\mathrm{arctanh}(e^{as})$, and

$$\tan(\psi) = (\tanh(\frac{as}{2}))^{\mp} \cdot y(t). \tag{78}$$

In case (ii), if $s > 0$ and $y(t) = \pm 1$ then $\psi = \pm\frac{1}{2}(as + \pi)$, mod$\pi$, and if $s < 0$ and $y(t) = \pm 1$, then $\psi = \pm\frac{1}{2}as$, modπ.

It remains to integrate the second equation of the Lax pair (67) using solutions for both $Q(s)$ and $\tan(\psi)$. The first case (i) is not hard and will be shown explicitly here. The others can be done, and more complicated cases are considered in the Appendix.

(i) Consider $Q(s) = s^{-1}$ and $\tan(\psi) = s^{-1} \cdot y(t)$. The second equation in (67) simplifies considerably to $y_t = -1$, therefore,

$$y(t) = -(t + \sigma), \qquad \tan(\psi) = -\frac{(t + \sigma)}{s}. \tag{79}$$

For $Q(s) = -s^{-1}$ and $\tan(\psi) = s \cdot y(t)$, the second equation of (67) becomes $y_t = -y^2$, therefore,

$$y(t) = \frac{1}{t + \sigma}, \qquad \tan(\psi) = \frac{s}{t + \sigma.} \tag{80}$$

8. A third-order equation for H and fundamental forms

Since $\xi_{12} = (\log e(s))' dt$, using Eq. (60) ω_{12} can be written as

$$\omega_{12} = \xi_{12} - d\psi = (\log e(s))' \, dt - d\psi. \tag{81}$$

Using Eqs. (14) and (64) for α_1, it follows that

$$d\log(J) = Q(\cos(2\psi)\, ds - \sin(2\psi)\, dt) - 2 * (\psi_t\, dt + \psi_s\, ds) + 2 * (\log(e(s)))' \, dt.$$

when ω_i are put in the s,t coordinates, using $*\omega_1 = \omega_2$, it can be stated that $*ds = dt$ and $*dt = -ds$. Consequently, $d\log(J)$ simplifies to

$$d\log(J) = (Q\cos(2\psi) + 2\psi_t - 2(\log(e(s)))')\, ds + (-Q\sin(2\psi) - 2\psi_s)\, dt. \tag{82}$$

First-order system (67) permits this to be written using $e(s) = \sqrt{E(s)}$ as

$$(\log(J))' + (\log(E))' = (\log(Q))'. \tag{83}$$

Hence, there exists a constant τ independent of s such that $E \cdot J = \tau Q$ or

$$E = \tau \frac{Q}{J} = \tau \frac{Q^2}{H'}. \tag{84}$$

This result (84) for E is substituted into the Gauss equation $-((log(E))_{ss}+(log(E))_{tt})=2E(H^2-J^2)$ giving

$$(\log(E))'' = 2(\log(Q))''-(\log(H_s))'' = 2Q^2-\left(\frac{H''}{H'}\right)'. \tag{85}$$

Therefore, the Gauss equation transforms into a third-order differential equation in the s variable,

$$\left(\frac{H''}{H'}\right)' + 2\tau H = 2Q^2\left(1 + \tau\frac{H^2}{H'}\right). \tag{86}$$

Thus, a characterization of Bonnet surfaces is reached by means of the solutions to these equations. This equation determines the function $H(s)$ and after that the functions $J(s)$ and $E(s)$. Therefore, Bonnet surfaces have as first fundamental form the expression

$$I = E(s)(ds^2 + dt^2), \qquad E(s) = \tau\frac{Q^2(s)}{H'(s)}. \tag{87}$$

Since ψ is the angle from the principal axis e_1 to the s curve with t equals constant, the second fundamental form is given by

$$II = L\, ds^2 + 2M\, ds\, dt + N\, dt^2. \tag{88}$$

where the coefficients L, M, N are given by

$$\begin{aligned} L &= E(H + J\cos(2\psi)) = EH + \tau Q\cos(2\psi), \\ M &= -EJ\sin(2\psi) = -\tau Q\ \sin(2\psi), \\ N &= E(H-J\cos(2\psi)). \end{aligned} \tag{89}$$

Appendix

It is worth seeing how the second equation in (67) can be integrated for cases (*ii*) and (*iii*). Only the case $s > 0$ will be done with $Q(s)$ taken from Eq. (73).

(a) Differentiating $\tan(\psi)$ given in Eq. (77), we obtain that

$$\psi_t = \frac{\tan\left(\frac{as}{2}\right)}{\tan^2\left(\frac{as}{2}\right) + y^2} y_t(t).$$

The following identities are required to simplify the result,

$$\tan(as) = \frac{2\tan\left(\frac{as}{2}\right)}{1-\tan^2\left(\frac{as}{2}\right)}, \qquad \cos(2\psi) = \frac{\tan^2\left(\frac{as}{2}\right)-y^2}{\tan^2\left(\frac{as}{2}\right)+y^2}.$$

Substituting ψ_t into Eq. (67), we obtain

$$\frac{2\tan\left(\frac{as}{2}\right)}{\tan^2\left(\frac{as}{2}\right)+y^2}y_t = -a\cot(as) - \frac{a}{\sin(as)}\frac{\tan^2\left(\frac{as}{2}\right)-y^2}{\tan^2\left(\frac{as}{2}\right)+y^2}.$$

Simplifying this, we get

$$\frac{4}{a}y_t = -\frac{1}{2}\left(1-\tan^2\left(\frac{as}{2}\right)\right)-\frac{1}{2}\left(\cot^2\left(\frac{as}{2}\right)-1\right)y^2-\sec^2\left(\frac{as}{2}\right)+\csc^2\left(\frac{as}{2}\right)y^2.$$

This simplifies to the elementary equation,

$$y_t = \frac{a}{2}(y^2-1), \qquad y(t) = -\tanh\left(\frac{at}{2}+\eta\right).$$

Here η is an integration constant. To summarize then,

$$\tan(\psi) = \tanh\left(\frac{at}{2}+\eta\right)\cdot\tan\left(\frac{as+\pi}{2}\right).$$

(*b*) Consider now $s>0$ and take $Q(s)$ from the last line of Eq. (73). Differentiating $\tan(\psi)$ from (78), we get

$$\psi_t = \frac{\coth\left(\frac{as}{2}\right)}{1+\coth^2\left(\frac{as}{2}\right)y^2}y_t(t).$$

In this case, the following identities are needed,

$$\tanh(as) = \frac{2\tanh\left(\frac{as}{2}\right)}{1+\tanh^2\left(\frac{as}{2}\right)}, \qquad \cos(2\psi) = \frac{1-\coth^2\left(\frac{as}{2}\right)y^2}{1+\coth^2\left(\frac{as}{2}\right)y^2}.$$

Therefore, Eq. (67) becomes

$$2\frac{\coth\left(\frac{as}{2}\right)}{1+\coth^2\left(\frac{as}{2}\right)y^2}y_t = -a\coth(as) - \frac{a}{\sinh(as)}\frac{\tanh^2\left(\frac{as}{2}\right)-y^2}{\tanh^2\left(\frac{as}{2}\right)+y^2}.$$

This reduces to

$$-\frac{4}{a}y_t = \left(1+\tanh^2\left(\frac{as}{2}\right)+\operatorname{sech}^2\left(\frac{as}{2}\right)\right)+\left(\coth^2\left(\frac{as}{2}\right)+1-\operatorname{csch}^2\left(\frac{as}{2}\right)\right)y^2.$$

Simplifying and integrating, it has been found that

$$y_t = -\frac{a}{2}(1+y^2), \qquad y(t) = -\tan\left(\frac{at}{2}+\eta\right).$$

To summarize then, it has been shown that,

$$\tan(\psi) = \cot\left(\frac{at}{2}+\eta\right)\cdot\coth\left(\frac{as}{2}\right).$$

These results apply to the case $s > 0$ and similar results can be found for the case $s < 0$ as well.

MSCs: 53A05, 58A10, 53B05

Author details

Paul Bracken

Address all correspondence to: paul.bracken@utrgv.edu

Department of Mathematics, University of Texas, Edinburg, TX, USA

References

[1] Bobenko A I, Eitner U. Painlevé Equations in the Differential Geometry of Surfaces, Lecture Notes in Mathematics, vol. 1753, Springer-Verlag, Berlin; 2000.

[2] Darboux G. Theory of Surfaces, Part 3, Paris, p. 304; 1894.

[3] Chern S S. Surface Theory with Darboux and Bianchi, Miscallanea Mathematica, Springer-Verlag, Berlin, 1991; 59–69.

[4] Hopf H. Differential Geometry in the Large, Part II, Lecture Notes in Mathematics, vol. 1000, Springer, Berlin-Heidelberg; 1983.

[5] Bonnet O. Mémoire sur la theorie des surface applicables sur une surface donneé, J. l'École Pol., Paris, 1867; XLII Cahier, 72–92.

[6] Chern S S. Deformations of Surfaces Preserving Principal Curvatures, Differential Geometry and Complex Analysis, H. E. Rauch Memorial Volume, Springer Verlag, Berlin-Heidelberg, 1985; 155–163.

[7] Roussos I M, Hernandez G E. On the number of distinct isometric immersions of a Riemannian surface into R^3 with given mean curvature. Am. J. Math. 1990; **192**, 71–85.

[8] Roussos I M. Global results on Bonnet surfaces. J. Geometry, 1999; **65**, 151–168.

http://dx.doi.org/10.5772/67008

[9] Lawson H B, Tribuzy R. On the mean curvature function for compact surfaces. J. Diff. Geometry, 1981; **16**, 179–183.

[10] Colares A G, Kenmotsu K. Isometric deformations of surfaces in R^3 preserving the mean curvature function, Pacific J. Math. 1989; **136, 1**, 71–80.

[11] Chen X, Peng C K. Deformations of Surfaces Preserving Principal Curvatures, Lecture Notes in Math., vol. 1369, Springer Verlag, Berlin-Heidelberg, 1987; 63–73.

[12] Chern S S, Chen W H, Lam K S. Lectures in Differential Geometry, Series on University Mathematics, vol. 1, World Scientific, Singapore, 1999.

[13] Kenmotsu K. An intrinsic characterization of *H*-deformable surfaces, J. London Math. Soc., 1994; **49, 2**, 555–568.

[14] Bracken P. Cartan's theory of moving frames and an application to a theorem of Bonnet. Tensor, 2008; **70**, 261–274.

Section 3

Study of Types of Manifolds

Chapter 3

Sub-Manifolds of a Riemannian Manifold

Mehmet Atçeken, Ümit Yıldırım and Süleyman Dirik

Additional information is available at the end of the chapter

http://dx.doi.org/10.5772/65948

Abstract

In this chapter, we introduce the theory of sub-manifolds of a Riemannian manifold. The fundamental notations are given. The theory of sub-manifolds of an almost Riemannian product manifold is one of the most interesting topics in differential geometry. According to the behaviour of the tangent bundle of a sub-manifold, with respect to the action of almost Riemannian product structure of the ambient manifolds, we have three typical classes of sub-manifolds such as invariant sub-manifolds, anti-invariant sub-manifolds and semi-invariant sub-manifolds. In addition, slant, semi-slant and pseudo-slant sub-manifolds are introduced by many geometers.

Keywords: Riemannian product manifold, Riemannian product structure, integral manifold, a distribution on a manifold, real product space forms, a slant distribution

1. Introduction

Let $i : M \to \tilde{M}$ be an immersion of an n-dimensional manifold M into an m-dimensional Riemannian manifold $(\tilde{M}, \tilde{g})$. Denote by $g = i^{*}\tilde{g}$ the induced Riemannian metric on M. Thus, i become an isometric immersion and M is also a Riemannian manifold with the Riemannian metric $g(X, Y) = \tilde{g}(X, Y)$ for any vector fields X, Y in M. The Riemannian metric g on M is called the induced metric on M. In local components, $g_{ij} = g_{AB}B_j^B B_i^A$ with $g = g_{ji}dx^j dx^j$ and $\tilde{g} = g_{BA}dU^B dU^A$.

If a vector field ξ_p of $\tilde{M}$ at a point $p \in M$ satisfies

$$\tilde{g}(X_p, \xi_p) = 0 \tag{1}$$

for any vector X_p of M at p, then ξ_p is called a normal vector of M in $\tilde{M}$ at p. A unit normal vector field of M in $\tilde{M}$ is called a normal section on M [3].

By $T^{\perp}M$, we denote the vector bundle of all normal vectors of M in $\tilde{M}$. Then, the tangent bundle of $\tilde{M}$ is the direct sum of the tangent bundle TM of M and the normal bundle $T^{\perp}M$ of M in $\tilde{M}$, i.e.,

$$T\tilde{M} = TM \oplus T^{\perp}M. \tag{2}$$

We note that if the sub-manifold M is of codimension one in $\tilde{M}$ and they are both orientiable, we can always choose a normal section ξ on M, i.e.,

$$g(X, \xi) = 0, \; g(\xi, \xi) = 1, \tag{3}$$

where X is any arbitrary vector field on M.

By $\tilde{\nabla}$, denote the Riemannian connection on $\tilde{M}$ and we put

$$\tilde{\nabla}_X Y = \nabla_X Y + h(X, Y) \tag{4}$$

for any vector fields X, Y tangent to M, where $\nabla_X Y$ and $h(X, Y)$ are tangential and the normal components of $\tilde{\nabla}_X Y$, respectively. Formula (4) is called the Gauss formula for the sub-manifold M of a Riemannian manifold $(\tilde{M}, \tilde{g})$.

Proposition 1.1. ∇ is the Riemannian connection of the induced metric $g = i^* \tilde{g}$ on M and $h(X, Y)$ is a normal vector field over M, which is symmetric and bilinear in X and Y.

Proof: Let α and β be differentiable functions on M. Then, we have

$$\begin{aligned} \tilde{\nabla}_{\alpha X}(\beta Y) &= \alpha\{X(\beta)Y + \beta\tilde{\nabla}_X Y\} \\ &= \alpha\{X(\beta)Y + \beta\nabla_X Y + \beta h(X, Y)\} \\ \nabla_{\alpha X}\beta Y + h(\alpha X, \beta Y) &= \alpha\beta\nabla_X Y + \alpha X(\beta)Y + \alpha\beta h(X, Y) \end{aligned} \tag{5}$$

This implies that

$$\nabla_{\alpha X}(\beta Y) = \alpha X(\beta)Y + \alpha\beta\nabla_X Y \tag{6}$$

and

$$h(\alpha X, \beta Y) = \alpha\beta h(X, Y). \tag{7}$$

Eq. (6) shows that ∇ defines an affine connection on M and Eq. (4) shows that h is bilinear in X and Y since additivity is trivial [1].

Since the Riemannian connection $\tilde{\nabla}$ has no torsion, we have

$$0 = \tilde{\nabla}_X Y - \tilde{\nabla}_Y X - [X, Y] = \nabla_X Y + h(X, Y) - \nabla_X Y - h(Y, X) - [X, Y]. \tag{8}$$

By comparing the tangential and normal parts of the last equality, we obtain

$$\nabla_X Y - \nabla_Y X = [X, Y] \tag{9}$$

and

$$h(X, Y) = h(Y, X). \tag{10}$$

These equations show that ∇ has no torsion and h is a symmetric bilinear map. Since the metric $\tilde{g}$ is parallel, we can easily see that

$$\begin{aligned}(\nabla_X g)(Y, Z) &= (\tilde{\nabla}_X \tilde{g})(Y, Z) \\ &= \tilde{g}(\tilde{\nabla}_X Y, Z) + \tilde{g}(Y, \tilde{\nabla}_X Z) \\ &- \tilde{g}\left(\nabla_X Y + h(X, Y), Z\right) + \tilde{g}(Y, \nabla_X Z + h(X, Z)) \\ &= \tilde{g}(\nabla_X Y, Z) + \tilde{g}(Y, \nabla_X Z) \\ &= g(\nabla_X Y, Z) + g(Y, \nabla_X Z)\end{aligned} \tag{11}$$

for any vector fields X, Y, Z tangent to M, that is, ∇ is also the Riemannian connection of the induced metric g on M.

We recall h the second fundamental form of the sub-manifold M (or immersion i), which is defined by

$$h : \Gamma(TM) \times \Gamma(TM) \to \Gamma(T^{\perp}M). \tag{12}$$

If $h = 0$ identically, then sub-manifold M is said to be totally geodesic, where $\Gamma(T^{\perp}M)$ is the set of the differentiable vector fields on normal bundle of M.

Totally geodesic sub-manifolds are simplest sub-manifolds.

Definition 1.1. Let M be an n-dimensional sub-manifold of an m-dimensional Riemannian manifold $(\tilde{M}, \tilde{g})$. By h, we denote the second fundamental form of M in $\tilde{M}$.

$H = \frac{1}{n}\mathrm{trace}(h)$ is called the mean curvature vector of M in $\tilde{M}$. If $H = 0$, the sub-manifold is called minimal.

On the other hand, M is called pseudo-umbilical if there exists a function λ on M, such that

$$\tilde{g}\left(h(X, Y), H\right) = \lambda g(X, Y) \tag{13}$$

for any vector fields X, Y on M and M is called totally umbilical sub-manifold if

$$h(X, Y) = g(X, Y)H. \tag{14}$$

It is clear that every minimal sub-manifold is pseudo-umbilical with $\lambda = 0$. On the other hand, by a direct calculation, we can find $\lambda = \tilde{g}(H, H)$ for a pseudo-umbilical sub-manifold. So, every totally umbilical sub-manifold is a pseudo-umbilical and a totally umbilical sub-manifold is totally geodesic if and only if it is minimal [2].

Now, let M be a sub-manifold of a Riemannian manifold $(\tilde{M}, \tilde{g})$ and V be a normal vector field on M, X be a vector field on M. Then, we decompose

$$\tilde{\nabla}_X V = -A_V X + \nabla^{\perp}_X V, \tag{15}$$

where $A_V X$ and $\nabla^{\perp}_X V$ denote the tangential and the normal components of $\nabla^{\perp}_X V$, respectively. We can easily see that $A_V X$ and $\nabla^{\perp}_X V$ are both differentiable vector fields on M and normal bundle of M, respectively. Moreover, Eq. (15) is also called Weingarten formula.

Proposition 1.2. Let M be a sub-manifold of a Riemannian manifold $(\tilde{M}, \tilde{g})$. Then

(a) $A_V X$ is bilinear in vector fields V and X. Hence, $A_V X$ at point $p \in M$ depends only on vector fields V_p and X_p.

(b) For any normal vector field V on M, we have

$$g(A_V X, Y) = g\Big(h(X, Y), V\Big). \tag{16}$$

Proof: Let α and β be any two functions on M. Then, we have

$$\begin{aligned} \tilde{\nabla}_{\alpha X}(\beta V) &= \alpha \tilde{\nabla}_X(\beta V) \\ &= \alpha\{X(\beta)V + \beta \tilde{\nabla}_X V\} \\ -A_{\beta V}\alpha X + \nabla^{\perp}_{\alpha X}\beta V &= \alpha X(\beta)V - \alpha\beta A_V X + \alpha\beta \nabla^{\perp}_X V. \end{aligned} \tag{17}$$

This implies that

$$A_{\beta V}\alpha X = \alpha\beta A_V X \tag{18}$$

and

$$\nabla^{\perp}_{\alpha X}\beta V = \alpha X(\beta)V + \alpha\beta \nabla^{\perp}_X V. \tag{19}$$

Thus, $A_V X$ is bilinear in V and X. Additivity is trivial. On the other hand, since $g$is a Riemannian metric,

$$X\tilde{g}(Y, V) = 0, \tag{20}$$

for any $X, Y \in \Gamma(TM)$ and $V \in \Gamma(T^{\perp}M)$.

Eq. (12) implies that

$$\tilde{g}(\tilde{\nabla}_X Y, V) + \tilde{g}(Y, \tilde{\nabla}_X V) = 0. \tag{21}$$

By means of Eqs. (4) and (15), we obtain

$$\tilde{g}\left(h(X,Y),V\right)-g(A_VX,Y)=0. \tag{22}$$

The proof is completed [3].

Let M be a sub-manifold of a Riemannian manifold $(\tilde{M},\tilde{g})$, and h and A_V denote the second fundamental form and shape operator of M, respectively.

The covariant derivative of h and A_V is, respectively, defined by

$$(\tilde{\nabla}_Xh)(Y,Z)=\nabla_X^{\perp}h(Y,Z)-h(\nabla_XY,Z)-h(Y,\nabla_XZ) \tag{23}$$

and

$$(\nabla_XA)_VY=\nabla_X(A_VY)-A_{\nabla_X^{\perp}V}Y-A_V\nabla_XY \tag{24}$$

for any vector fields X,Y tangent to M and any vector field V normal to M. If $\nabla_Xh=0$ for all X, then the second fundamental form of M is said to be parallel, which is equivalent to $\nabla_XA=0$. By direct calculations, we get the relation

$$g\left((\nabla_Xh)(Y,Z),V\right)=g\left((\nabla_XA)_VY,Z\right). \tag{25}$$

Example 1.1. We consider the isometric immersion

$$\phi:\mathbb{R}^2\to\mathbb{R}^4, \tag{26}$$

$$\phi(x_1,x_2)=(x_1,\sqrt{x_1^2-1},x_2,\sqrt{x_2^2-1}) \tag{27}$$

we note that $M=\phi(\mathbb{R}^2)\subset\mathbb{R}^4$ is a two-dimensional sub-manifold of $\mathbb{R}^4$ and the tangent bundle is spanned by the vectors

$TM=S_p\left\{e_1=\left(\sqrt{x_1^2-1},x_1,0,0\right),e_2=\left(0,0,\sqrt{x_2^2-1},x_2\right)\right\}$ and the normal vector fields

$$T^{\perp}M=sp\left\{w_1=\left(-x_1,\sqrt{x_1^2-1},0,0),w_2=(0,0,-x_1,\sqrt{x_2^2-1}\right)\right\}. \tag{28}$$

By $\tilde{\nabla}$, we denote the Levi-Civita connection of $\mathbb{R}^4$, the coefficients of connection, are given by

$$\tilde{\nabla}_{e_1}e_1=\frac{2x_1\sqrt{x_1^2-1}}{2x_1^2-1}e_1-\frac{1}{2x_1^2-1}w_1, \tag{29}$$

$$\tilde{\nabla}_{e_2}e_2=\frac{2x_2\sqrt{x_2^2-1}}{2x_2^2-1}e_2-\frac{1}{2x_2^2-1}w_2 \tag{30}$$

and

$$\nabla_{e_2} e_1 = 0. \tag{31}$$

Thus, we have $h(e_1, e_1) = -\frac{1}{2x_1^2-1} w_1$, $h(e_2, e_2) = -\frac{1}{2x_2^2-1} w_2$ and $h(e_2, e_1) = 0$. The mean curvature vector of $M = \phi(\mathbb{R}^2)$ is given by

$$H = -\frac{1}{2}(w_1 + w_2). \tag{32}$$

Furthermore, by using Eq. (16), we obtain

$$\begin{aligned} g(A_{w_1} e_1, e_1) &= g\Big(h(e_1, e_1), w_1\Big) = -\frac{1}{2x_1^2-1}(x_1^2 + x_1^2-1) = -1, \\ g(A_{w_1} e_2, e_2) &= g\Big(h(e_2, e_2), w_1\Big) = -\frac{1}{2x_2^2-1} g(w_1, w_2) = 0, \\ g(A_{w_1} e_1, e_2) &= 0, \end{aligned} \tag{33}$$

and

$$\begin{aligned} g(A_{w_2} e_1, e_1) &= g\Big(h(e_1, e_1), w_2\Big) = 0, \\ g(A_{w_2} e_1, e_2) &= 0, g(A_{w_2} e_2, e_2) = 1. \end{aligned} \tag{34}$$

Thus, we have

$$A_{w_1} = \begin{pmatrix} -1 & 0 \\ 0 & 0 \end{pmatrix} \text{ and } A_{w_2} = \begin{pmatrix} 0 & 0 \\ 0 & -1 \end{pmatrix}. \tag{35}$$

Now, let M be a sub-manifold of a Riemannian manifold $(\tilde{M}, g)$, $\tilde{R}$ and R be the Riemannian curvature tensors of $\tilde{M}$ and M, respectively. From then the Gauss and Weingarten formulas, we have

$$\begin{aligned} \tilde{R}(X, Y)Z &= \tilde{\nabla}_X \tilde{\nabla}_Y Z - \tilde{\nabla}_Y \tilde{\nabla}_X Z - \tilde{\nabla}_{[X,Y]} Z \\ &= \tilde{\nabla}_X\Big(\nabla_Y Z + h(Y, Z)\Big) - \tilde{\nabla}_Y\Big(\nabla_X Z + h(X, Z)\Big) - \nabla_{[X,Y]} Z - h([X, Y], Z) \\ &= \tilde{\nabla}_X \nabla_Y Z + \tilde{\nabla}_X h(Y, Z) - \tilde{\nabla}_Y \nabla_X Z - \tilde{\nabla}_Y h(X, Z) - \nabla_{[X,Y]} Z - h(\nabla_X Y, Z) + h(\nabla_Y X, Z) \\ &= \nabla_X \nabla_Y Z - \nabla_Y \nabla_X Z + h(X, \nabla_Y Z) - h(\nabla_X Z, Y) + \nabla_X^{\perp} h(Y, Z) \\ &\quad - A_{h(Y,Z)} X - \nabla_Y^{\perp} h(X, Z) + A_{h(X,Z)} Y - \nabla_{[X,Y]} Z - h(\nabla_X Y, Z) + h(\nabla_Y X, Z) \\ &= \nabla_X \nabla_Y Z - \nabla_Y \nabla_X Z - \nabla_{[X,Y]} Z + \nabla_X^{\perp} h(Y, Z) - h(\nabla_X Y, Z) \\ &\quad - h(Y, \nabla_X Z) - \nabla_Y^{\perp} h(X, Z) + h(\nabla_Y X, Z) + h(\nabla_Y Z, X) \\ &\quad + A_{h(X,Z)} Y - A_{h(Y,Z)} X \\ &= R(X, Y)Z + (\nabla_X h)(Y, Z) - (\nabla_Y h)(X, Z) + A_{h(X,Z)} Y - A_{h(Y,Z)} X \end{aligned} \tag{36}$$

from which

$$\tilde{R}(X,Y)Z = R(X,Y)Z + A_{h(X,Z)}Y - A_{h(Y,Z)}X + (\nabla_X h)(Y,Z) - (\nabla_Y h)(X,Z), \tag{37}$$

for any vector fields X, Y and Z tangent to M. For any vector field W tangent to M, Eq. (37) gives the Gauss equation

$$g\left(\tilde{R}(X,Y)Z,W\right) = g\left(R(X,Y)Z,W\right) + g\left(h(Y,W),h(X,Z)\right) - g\left(h(Y,Z),h(X,W)\right). \tag{38}$$

On the other hand, the normal component of Eq. (37) is called equation of Codazzi, which is given by

$$\left(\tilde{R}(X,Y)Z\right)^{\perp} = (\nabla_X h)(Y,Z) - (\nabla_Y h)(X,Z). \tag{39}$$

If the Codazzi equation vanishes identically, then sub-manifold M is said to be curvature-invariant sub-manifold [4].

In particular, if $\tilde{M}$ is of constant curvature, $\tilde{R}(X,Y)Z$ is tangent to M, that is, sub-manifold is curvature-invariant. Whereas, in Kenmotsu space forms, and Sasakian space forms, this not true.

Next, we will define the curvature tensor $R^{\perp}$ of the normal bundle of the sub-manifold M by

$$R^{\perp}(X,Y)V = \nabla^{\perp}_X \nabla^{\perp}_Y V - \nabla^{\perp}_Y \nabla^{\perp}_X V - \nabla^{\perp}_{[X,Y]} V \tag{40}$$

for any vector fields X, Y tangent to sub-manifold M, and any vector field V normal to M. From the Gauss and Weingarten formulas, we have

$$\begin{aligned}
\tilde{R}(X,Y)V &= \tilde{\nabla}_X \tilde{\nabla}_Y V - \tilde{\nabla}_Y \tilde{\nabla}_X V - \tilde{\nabla}_{[X,Y]} V \\
&= \tilde{\nabla}_X(-A_V Y + \nabla^{\perp}_Y V) - \tilde{\nabla}_Y(-A_V X + \nabla^{\perp}_X V) + A_V[X,Y] - \nabla^{\perp}_{[X,Y]} V \\
&= -\tilde{\nabla}_X A_V Y + \tilde{\nabla}_Y A_V X + \tilde{\nabla}_X \nabla^{\perp}_Y V - \tilde{\nabla}_Y \nabla^{\perp}_X V + A_V[X,Y] - \nabla^{\perp}_{[X,Y]} V \\
&= -\nabla_X A_V Y - h(X, A_V Y) + \nabla_Y A_V X + h(Y, A_V X) \\
&\quad + \nabla^{\perp}_X \nabla^{\perp}_Y V - \nabla^{\perp}_Y \nabla^{\perp}_X V - A_{\nabla^{\perp}_Y V} X + A_{\nabla^{\perp}_X V} Y + A_V[X,Y] - \nabla^{\perp}_{[X,Y]} V \\
&= \nabla^{\perp}_X \nabla^{\perp}_Y V - \nabla^{\perp}_Y \nabla^{\perp}_X V - \nabla^{\perp}_{[X,Y]} V - A_{\nabla^{\perp}_Y V} X + A_{\nabla^{\perp}_X V} Y + A_V[X,Y] \\
&\quad - \nabla_X A_V Y + \nabla_Y A_V X - h(X, A_V Y) + h(Y, A_V X) \\
&= R^{\perp}(X,Y)V + h(A_V X, Y) - h(X, A_V Y) - (\nabla_X A)_V Y + (\nabla_Y A)_V X.
\end{aligned} \tag{41}$$

For any normal vector U to M, we obtain

$$\begin{aligned} g\left(\tilde{R}\,(X,Y)V,U\right) &= g\left(R^{\perp}(X,Y)V,U\right) + g\left(h(A_V X,Y),U\right) - g\left(h(X,A_V Y),U\right) \\ &= g\left(R^{\perp}(X,Y)V,U\right) + g(A_U Y,A_V X) - g(A_V Y,A_U X) \\ &= g\left(R^{\perp}(X,Y)V,U\right) + g(A_V A_U Y,X) - g(A_U A_V Y,X) \end{aligned} \tag{42}$$

Since $[A_U, A_V] = A_U A_V - A_V A_U$, Eq. (42) implies

$$g\left(\tilde{R}(X,Y)V,U\right) = g\left(R^{\perp}(X,Y)V,U\right) + g([A_U,A_V]Y,X). \tag{43}$$

Eq. (43) is also called the Ricci equation.

If $R^{\perp} = 0$, then the normal connection of M is said to be flat [2].

When $\left(\tilde{R}(X,Y)V\right)^{\perp} = 0$, the normal connection of the sub-manifold M is flat if and only if the second fundamental form M is commutative, i.e. $[A_U, A_V] = 0$ for all U, V. If the ambient space $\tilde{M}$ is real space form, then $\left(\tilde{R}(X,Y)V\right)^{\perp} = 0$ and hence the normal connection of M is flat if and only if the second fundamental form is commutative. If $\tilde{R}(X,Y)Z$ tangent to M, then equation of codazzi Eq. (37) reduces to

$$(\nabla_X h)(Y,Z) = (\nabla_Y h)(X,Z) \tag{44}$$

which is equivalent to

$$(\nabla_X A)_V Y = (\nabla_Y A)_V X. \tag{45}$$

On the other hand, if the ambient space $\tilde{M}$ is a space of constant curvature c, then we have

$$\tilde{R}(X,Y)Z = c\{g(Y,Z)X - g(X,Z)Y\} \tag{46}$$

for any vector fields X, Y and Z on $\tilde{M}$.

Since $\tilde{R}(X,Y)Z$ is tangent to M, the equation of Gauss and the equation of Ricci reduce to

$$\begin{aligned} g\left(R(X,Y)Z,W\right) &= c\{g(Y,Z)g(X,W) - g(X,Z)g(Y,W)\} \\ &\quad + g\left(h(Y,Z),h(X,W)\right) - g\left(h(Y,W),h(X,Z)\right) \end{aligned} \tag{47}$$

and

$$g\left(R^{\perp}(X,Y)V,U\right) = g([A_U,A_V]X,Y), \tag{48}$$

respectively.

http://dx.doi.org/10.5772/65948

Proposition 1.3. A totally umbilical sub-manifold M in a real space form $\tilde{M}$ of constant curvature c is also of constant curvature.

Proof: Since M is a totally umbilical sub-manifold of $\tilde{M}$ of constant curvature c, by using Eqs. (14) and (46), we have

$$\begin{aligned} g\left(R(X,Y)Z,W\right) &= c\{g(Y,Z)g(X,W)-g(X,Z)g(Y,W)\} \\ &\quad +g(H,H)\{g(Y,Z)g(X,W)-g(X,Z)g(Y,W)\} \\ &= \{c+g(H,H)\}\{g(Y,Z)g(X,W)-g(X,Z)g(Y,W)\}. \end{aligned} \tag{49}$$

This shows that the sub-manifold M is of constant curvature $c+\|H^2\|$ for $n>2$. If $n=2$, $\|H\|=$ constant follows from the equation of Codazzi [3].

This proves the proposition.

On the other hand, for any orthonormal basis $\{e_a\}$ of normal space, we have

$$\begin{aligned} g(Y,Z)g(X,W)-g(X,Z)g(Y,W) &= \sum_a\Big[g\left(h(Y,Z),e_a\right)g\left(h(X,W),e_a\right) \\ &\quad -g\left(h(X,Z),e_a\right)g\left(h(Y,W),e_a\right)\Big] \\ &= \sum_a g(A_{e_a}Y,Z)g(A_{e_a}X,W)-g(A_{e_a}X,Z)g(A_{e_a}Y,W) \end{aligned} \tag{50}$$

Thus, Eq. (45) can be rewritten as

$$\begin{aligned} g\left(R(X,Y)Z,W\right) &= c\{g(Y,Z)g(X,W)-g(X,Z)g(Y,W)\} \\ &\quad +\sum_a[g(A_{e_a}Y,Z)g(A_{e_a}X,W)-g(A_{e_a}X,Z)g(A_{e_a}Y,W)] \end{aligned} \tag{51}$$

By using A_{e_a}, we can construct a similar equation to Eq. (47) for Eq. (23).

Now, let S- be the Ricci tensor of M. Then, Eq. (47) gives us

$$S(X,Y) = c\{ng(X,Y)-g(e_i,X)g(e_i,Y)\} \tag{52}$$

$$\begin{aligned} &+\sum_{e_a}[g(A_{e_a}e_i,e_i)g(A_{e_a}X,Y)-g(A_{e_a}X,e_i)g(A_{e_a}e_i,Y)] \\ &= c(n-1)g(X,Y)+\sum_{e_a}[Tr(A_{e_a})g(A_{e_a}X,Y)-g(A_{e_a}X,A_{e_a}Y)], \end{aligned} \tag{53}$$

where $\{e_1,e_2,\ldots,e_n\}$ are orthonormal basis of M.

Therefore, the scalar curvature r of sub-manifold M is given by

$$r = cn(n-1)\sum_{e_a} Tr^2(A_{e_a}) - \sum_{e_a} Tr(A_{e_a})^2 \tag{54}$$

$\sum_{e_a} Tr(A_{e_a})^2$ is the square of the length of the second fundamental form of M, which is denoted by $|A_{e_a}|^2$. Thus, we also have

$$\| h^2 \| = \sum_{i,j=1}^{n} g\left(h(e_i, e_j), h(e_i, e_j)\right) = \| A^2 \|. \tag{55}$$

2. Distribution on a manifold

An m-dimensional distribution on a manifold $\tilde{M}$ is a mapping $\mathcal{D}$ defined on $\tilde{M}$, which assignes to each point p of $\tilde{M}$ an m-dimensional linear subspace $\mathcal{D}_p$ of $T_{\tilde{M}}(p)$. A vector field X on $\tilde{M}$ belongs to $\mathcal{D}$ if we have $X_p \in \mathcal{D}_p$ for each $p \in \tilde{M}$. When this happens, we write $X \in \Gamma(\mathcal{D})$. The distribution $\mathcal{D}$ is said to be differentiable if for any $p \in \tilde{M}$, there exist m-differentiable linearly independent vector fields $X_j \in \Gamma(\mathcal{D})$ in a neighbordhood of p.

The distribution $\mathcal{D}$ is said to be involutive if for all vector fields $X, Y \in \Gamma(\mathcal{D})$ we have $[X, Y] \in \Gamma(\mathcal{D})$. A sub-manifold M of $\tilde{M}$ is said to be an integral manifold of $\mathcal{D}$ if for every point $p \in M$, $\mathcal{D}_p$ coincides with the tangent space to M at p. If there exists no integral manifold of $\mathcal{D}$ which contains M, then M is called a maximal integral manifold or a leaf of $\mathcal{D}$. The distribution $\mathcal{D}$ is said to be integrable if for every $p \in \tilde{M}$, there exists an integral manifold of $\mathcal{D}$ containing p [2].

Let $\tilde{\nabla}$ and distribution be a linear connection on $\tilde{M}$, respectively. The distribution $\mathcal{D}$ is said to be parallel with respect to $\tilde{M}$, if we have

$$\tilde{\nabla}_X Y \in \Gamma(\mathcal{D}) \text{ for all } X \in \Gamma(T\tilde{M}) \text{ and } Y \in \Gamma(\mathcal{D}) \tag{56}$$

Now, let $(\tilde{M}, \tilde{g})$ be Riemannian manifold and $\mathcal{D}$ be a distribution on $\tilde{M}$. We suppose $\tilde{M}$ is endowed with two complementary distribution $\mathcal{D}$ and $\mathcal{D}^\perp$, i.e., we have $T\tilde{M} = \mathcal{D} \oplus \mathcal{D}^\perp$. Denoted by P and Q the projections of $T\tilde{M}$ to $\mathcal{D}$ and $\mathcal{D}^\perp$, respectively.

Theorem 2.1. All the linear connections with respect to which both distributions $\mathcal{D}$ and $\mathcal{D}^\perp$ are parallel, are given by

$$\nabla_X Y = P\nabla'_X PY + Q\nabla'_X QY + PS(X, PY) + QS(X, QY) \tag{57}$$

for any $X, Y \in \Gamma(T\tilde{M})$, where ∇' and S are, respectively, an arbitrary linear connection and arbitrary tensor field of type $(1, 2)$ on $\tilde{M}$.

Proof: Suppose ∇' is an arbitrary linear connection on $\tilde{M}$. Then, any linear connection ∇ on $\tilde{M}$ is given by

http://dx.doi.org/10.5772/65948

$$\nabla_X Y = \nabla'_X Y + S(X, Y) \tag{58}$$

for any $X, Y \in \Gamma(T\tilde{M})$. We can put

$$X = PX + QX \tag{59}$$

for any $X \in \Gamma(T\tilde{M})$. Then, we have

$$\begin{aligned}\nabla_X Y &= \nabla_X(PY + QY) = \nabla_X PY + \nabla_X QY = \nabla'_X PY + S(X, PY)\\ &+\nabla'_X QY + S(X, QY) = \mathrm{P}\nabla'_X PY + Q\nabla'_X PY + PS(X, PY) + QS(X, PY)\\ &+P\nabla'_X QY + Q\nabla'_X QY + PS(X, QY) + QS(X, QY)\end{aligned} \tag{60}$$

for any $X, Y \in \Gamma(T\tilde{M})$.

The distributions $\mathcal{D}$ and $\mathcal{D}$ are both parallel with respect to ∇ if and only if we have

$$\phi(\nabla_X PY) = 0 \mathrm{and} \mathrm{P}(\nabla_X QY) = 0. \tag{61}$$

From Eqs. (58) and (61), it follows that $\mathcal{D}$ and $\mathcal{D}^{\perp}$ are parallel with respect to ∇ if and only if

$$\mathrm{Q}\nabla'_X PY + QS(X, PY) = 0 \text{ and } \mathrm{P}\nabla'_X QY + PS(X, QY) = 0. \tag{62}$$

Thus, Eqs. (58) and (62) give us Eq. (57).

Next, by means of the projections P and Q, we define a tensor field F of type $(1, 1)$ on $\tilde{M}$ by

$$FX = PX\text{–}QX \tag{63}$$

for any $X \in \Gamma(T\tilde{M})$. By a direct calculation, it follows that $F^2 = I$. Thus, we say that F defines an almost product structure on $\tilde{M}$. The covariant derivative of F is defined by

$$(\nabla_X F)Y = \nabla_X FY\text{–}\mathrm{F}\nabla_X Y \tag{64}$$

for all $X, Y \in \Gamma(T\tilde{M})$. We say that the almost product structure F is parallel with respect to the connection ∇, if we have $\nabla_X F = 0$. In this case, F is called the Riemannian product structure [2].

Theorem 2.2. Let $(\tilde{M}, \tilde{g})$ be a Riemannian manifold and $\mathcal{D}$, $\mathcal{D}^{\perp}$ be orthogonal distributions on $\tilde{M}$ such that $T\tilde{M} = \mathcal{D} \oplus \mathcal{D}^{\perp}$. Both distributions $\mathcal{D}$ and $\mathcal{D}^{\perp}$ are parallel with respect to ∇ if and only if F is a Riemannian product structure.

Proof: For any $X, Y \in \Gamma(T\tilde{M})$, we can write

$$\tilde{\nabla}_Y PX = \tilde{\nabla}_{PY} PX + \tilde{\nabla}_{QY} PX \tag{65}$$

and

$$\tilde{\nabla}_Y X = \tilde{\nabla}_{PY} PX + \tilde{\nabla}_{PY} QX + \tilde{\nabla}_{QY} PX + \tilde{\nabla}_{QY} QX, \tag{66}$$

from which

$$g(\tilde{\nabla}_{QY} PX, QZ) = QYg(PX, QZ) - g(\nabla_{QY} QZ, PX) = 0 - g(\tilde{\nabla}_{QY} QZ, PX) = 0, \tag{67}$$

that is, $\nabla_{QY} PX \in \Gamma(\mathcal{D})$ and so $P\tilde{\nabla}_{QY} PX = \tilde{\nabla}_{QY} PX$,

$$Q\tilde{\nabla}_{QY} PX = 0. \tag{68}$$

In the same way, we obtain

$$g(\tilde{\nabla}_{PY} QX, PZ) = PYg(QX, PZ) - g(QX, \tilde{\nabla}_{PY} PZ) = 0, \tag{69}$$

which implies that

$$P\tilde{\nabla}_{PY} QX = 0 \text{ and } Q\tilde{\nabla}_{PY} QX = \tilde{\nabla}_{PY} QX. \tag{70}$$

From Eqs. (66), (68) and (70), it follows that

$$P\tilde{\nabla}_Y X = \tilde{\nabla}_{PY} PX + \tilde{\nabla}_{QY} PX. \tag{71}$$

By using Eqs. (64) and (71), we obtain

$$(\tilde{\nabla}_Y P)X = \tilde{\nabla}_Y PX - P\tilde{\nabla}_Y X = \tilde{\nabla}_{PY} PX + \tilde{\nabla}_{QY} PX - \tilde{\nabla}_{PY} PX - \tilde{\nabla}_{QY} PX = 0. \tag{72}$$

In the same way, we can find $\tilde{\nabla} Q = 0$. Thus, we obtain

$$\tilde{\nabla} F = \tilde{\nabla}(P - Q) = 0. \tag{73}$$

This proves our assertion [2].

Theorem 2.3. Both distributions $\mathcal{D}$ and $\mathcal{D}^{\perp}$ are parallel with respect to Levi-Civita connection ∇ if and only if they are integrable and their leaves are totally geodesic in $\tilde{M}$.

Proof: Let us assume both distributions $\mathcal{D}$ and $\mathcal{D}^{\perp}$ are parallel. Since ∇ is a torsion free linear connection, we have

$$[X, Y] = \nabla_X Y - \nabla_Y X \in \Gamma(\mathcal{D}), \text{for any } X, Y \in \Gamma(\mathcal{D}) \tag{74}$$

and

$$[U, V] = \nabla_U V - \nabla_V U \in \Gamma(\mathcal{D}^{\perp}), \text{for any } U, V \in \Gamma(\mathcal{D}^{\perp}) \tag{75}$$

Thus, $\mathcal{D}$ and $\mathcal{D}^{\perp}$ are integrable distributions. Now, let M be a leaf of $\mathcal{D}$ and denote by h the second fundamental form of the immersion of M in $\tilde{M}$. Then by the Gauss formula, we have

$$\nabla_X Y = \nabla'_X Y + h(X, Y) \tag{76}$$

for any $X, Y \in \Gamma(\mathcal{D})$, where ∇' denote the Levi-Civita connection on M. Since $\mathcal{D}$ is parallel from Eq. (76) we conclude $h = 0$, that is, M is totally in $\tilde{M}$. In the same way, it follows that each leaf of $\mathcal{D}^{\perp}$ is totally geodesic in $\tilde{M}$.

Conversely, suppose $\mathcal{D}$ and $\mathcal{D}^{\perp}$ be integrable and their leaves are totally geodesic in $\tilde{M}$. Then by using Eq. (4), we have

$$\nabla_X Y \in \Gamma(\mathcal{D}) \text{ for any } X, Y \in \Gamma(\mathcal{D}) \tag{77}$$

and

$$\nabla_U V \in \Gamma(\mathcal{D}^{\perp}) \text{ for any } U, V \in \Gamma(\mathcal{D}^{\perp}). \tag{78}$$

Since g is a Riemannian metric tensor, we obtain

$$g(\nabla_U Y, V) = -g(Y, \nabla_U V) = 0 \tag{79}$$

and

$$g(\nabla_X V, Y) = -g(V, \nabla_X Y) = 0 \tag{80}$$

for any $X, Y \in \Gamma(\mathcal{D})$ and $U, V \in \Gamma(\mathcal{D}^{\perp})$. Thus, both distributions $\mathcal{D}$ and $\mathcal{D}^{\perp}$ are parallel on $\tilde{M}$.

3. Locally decomposable Riemannian manifolds

Let $(\tilde{M}, \tilde{g})$ be n–dimensional Riemannian manifold and F be a tensor $(1,1)$–type on $\tilde{M}$ such that $F^2 = I$, $F \neq \mp I$.

If the Riemannian metric tensor $\tilde{g}$ satisfying

$$\tilde{g}(X, Y) = \tilde{g}(FX, FY) \tag{81}$$

for any $X, Y \in \Gamma(T\tilde{M})$ then $\tilde{M}$ is called almost Riemannian product manifold and F is said to be almost Riemannian product structure. If F is parallel, that is, $(\tilde{\nabla}_X F)Y = 0$, then $\tilde{M}$ is said to be locally decomposable Riemannian manifold.

Now, let $\tilde{M}$ be an almost Riemannian product manifold. We put

$$P = \frac{1}{2}(I + F), \; Q = \frac{1}{2}(I - F). \tag{82}$$

Then, we have

$$P + Q = I, \quad P^2 = P, \quad Q^2 = Q, \quad PQ = QP = 0 \quad \text{and} \quad F = P - Q. \tag{83}$$

Thus, P and Q define two complementary distributions P and Q globally. Since $F^2 = I$, we easily see that the eigenvalues of F are 1 and −1. An eigenvector corresponding to the eigenvalue 1 is in P and an eigenvector corresponding to −1 is in Q. If F has eigenvalue 1 of multiplicity P and eigenvalue −1 of multiplicity q, then the dimension of P is p and that of Q is q. Conversely, if there exist in $\tilde{M}$ two globally complementary distributions P and Q of dimension p and q, respectively. Then, we can define an almost Riemannian product structure F on $\tilde{M}$ by $\tilde{M}$ by $F = P - Q$ [7].

Let $(\tilde{M}, \tilde{g}, F)$ be a locally decomposable Riemannian manifold and we denote the integral manifolds of the distributions P and Q by M^p and M^q, respectively. Then we can write $\tilde{M} = M^p X M^q$, $(p, q > 2)$. Also, we denote the components of the Riemannian curvature R of $\tilde{M}$ by R_{dcba}, $1 \leq a, b, c, d \leq n = p + q$.

Now, we suppose that the two components are both of constant curvature λ and μ. Then, we have

$$R_{dcba} = \lambda\{g_{da}g_{cb} - g_{ca}g_{db}\} \tag{84}$$

and

$$R_{zyxw} = \mu\{g_{zw}g_{yx} - g_{yw}g_{zx}\}. \tag{85}$$

Then, the above equations may also be written in the form

$$\begin{aligned} R_{kjih} = & \frac{1}{4}(\lambda + \mu)\{(g_{kh}g_{ji} - g_{jh}g_{ki}) + (F_{kh}F_{ji} - F_{jh}F_{ki})\} \\ & + \frac{1}{4}(\lambda - \mu)\{(F_{kh}g_{ji} - F_{jh}g_{ki}) + (g_{kh}F_{ji} - g_{jh}F_{ki})\}. \end{aligned} \tag{86}$$

Conversely, suppose that the curvature tensor of a locally decomposable Riemannian manifold has the form

$$\begin{aligned} R_{kjih} = & a\{(g_{kh}g_{ji} - g_{jh}g_{ki}) + (F_{kh}F_{ji} - F_{jh}F_{ki})\} \\ & + b\{(F_{kh}g_{ji} - F_{jh}g_{ki}) + (g_{kh}F_{ji} - g_{jh}F_{ki})\}. \end{aligned} \tag{87}$$

Then, we have

$$R_{cdba} = 2(a + b)\{g_{da}g_{cb} - g_{ca}g_{db}\} \tag{88}$$

and

$$R_{zyxw} = 2(a - b)\{g_{zw}g_{yx} - g_{yw}g_{zx}\}. \tag{89}$$

Let $\tilde{M}$ be an m−dimensional almost Riemannian product manifold with the Riemannian structure $(F, \tilde{g})$ and M be an n−dimensional sub-manifold of $\tilde{M}$. For any vector field X tangent to M, we put

$$FX = fX + wX, \tag{90}$$

where fX and wX denote the tangential and normal components of FX, with respect to M, respectively. In the same way, for $V \in \Gamma(T^{\perp}M)$, we also put

$$FV = BV + CV, \tag{91}$$

where BV and CV denote the tangential and normal components of FV, respectively.

Then, we have

$$f^2 + Bw = I, Cw + wf = 0 \tag{92}$$

and

$$fB + BC = 0, wB + C^2 = I. \tag{93}$$

On the other hand, we can easily see that

$$g(X, fY) = g(fX, Y) \tag{94}$$

and

$$g(X, Y) = g(fX, fY) + g(wX, wY) \tag{95}$$

for any $X, Y \in \Gamma(TM)$ [6].

If $wX = 0$ for all $X \in \Gamma(TM)$, then M is said to be invariant sub-manifold in $\tilde{M}$, i.e., $F(T_M(p)) \subset T_M(p)$ for each $p \in M$. In this case, $f^2 = I$ and $g(fX, fY) = g(X, Y)$. Thus, (f, g) defines an almost product Riemannian on M.

Conversely, (f, g) is an almost product Riemannian structure on M, the $w = 0$ and hence M is an invariant sub-manifold in $\tilde{M}$.

Consequently, we can give the following theorem [7].

Theorem 3.1. Let M be a sub-manifold of an almost Riemannian product manifold $\tilde{M}$ with almost Riemannian product structure $(F, \tilde{g})$. The induced structure (f, g) on M is an almost Riemannian product structure if and only if M is an invariant sub-manifold of $\tilde{M}$.

Definition 3.1. Let M be a sub-manifold of an almost Riemannian product $\tilde{M}$ with almost product Riemannian structure $(F, \tilde{g})$. For each non-zero vector $X_p \in T_M(p)$ at $p \in M$, we denote the slant angle between FX_p and $T_M(p)$ by $\theta(p)$. Then M said to be slant sub-manifold if the angle $\theta(p)$ is constant, i.e., it is independent of the choice of $p \in M$ and $X_p \in T_M(p)$ [5].

Thus, invariant and anti-invariant immersions are slant immersions with slant angle $\theta = 0$ and $\theta = \frac{\pi}{2}$, respectively. A proper slant immersion is neither invariant nor anti-invariant.

Theorem 3.2. Let M be a sub-manifold of an almost Riemannian product manifold $\tilde{M}$ with almost product Riemannian structure $(F, \tilde{g})$. M is a slant sub-manifold if and only if there exists a constant $\lambda \in (0, 1)$, such tha

$$f^2 = \lambda I. \tag{96}$$

Furthermore, if the slant angle is θ, then it satisfies $\lambda = \cos^2\theta$ [9].

Definition 3.2. Let M be a sub-manifold of an almost Riemannian product manifold $\tilde{M}$ with almost Riemannian product structure $(F, \tilde{g})$. M is said to be semi-slant sub-manifold if there exist distributions $\mathcal{D}^\theta$ and $\mathcal{D}^T$ on M such that

(i) TM has the orthogonal direct decomposition $TM = \mathcal{D} \oplus \mathcal{D}^T$.

(ii) The distribution $\mathcal{D}^\theta$ is a slant distribution with slant angle θ.

(iii) The distribution $\mathcal{D}^T$ is an invariant distribution, .e., $F(\mathcal{D}_\mathrm{T}) \subseteq \mathcal{D}^T$.

In a semi-slant sub-manifold, if $\theta = \frac{\pi}{2}$, then semi-slant sub-manifold is called semi-invariant sub-manifold [8].

Example 3.1. Now, let us consider an immersed sub-manifold M in $\mathbb{R}^7$ given by the equations

$$x_1^2 + x_2^2 = x_5^2 + x_6^2, x_3 + x_4 = 0. \tag{97}$$

By direct calculations, it is easy to check that the tangent bundle of M is spanned by the vectors

$$\begin{gathered} z_1 = \cos\theta \frac{\partial}{\partial x_1} + \sin\theta \frac{\partial}{\partial x_2} + \cos\beta \frac{\partial}{\partial x_5} + \sin\beta \frac{\partial}{\partial x_6} \\ z_2 = -u\sin\theta \frac{\partial}{\partial x_1} + u\cos\theta \frac{\partial}{\partial x_2}, z_3 = \frac{\partial}{\partial x_3} - \frac{\partial}{\partial x_4}, \\ z_4 = -u\sin\beta \frac{\partial}{\partial x_5} + u\cos\beta \frac{\partial}{\partial x_6}, z_5 = \frac{\partial}{\partial x_7}, \end{gathered} \tag{98}$$

where θ, β and u denote arbitrary parameters.

For the coordinate system of $\mathbb{R}^7 = \{(x_1, x_2, x_3, x_4, x_5, x_6, x_7) | x_i \in \mathbb{R}, 1 \leq i \leq 7\}$, we define the almost product Riemannian structure F as follows:

$$F\left(\frac{\partial}{\partial x_i}\right) = \frac{\partial}{\partial x_i}, F\left(\frac{\partial}{\partial x_j}\right) = \frac{\partial}{\partial x_j}, 1 \leq i \leq 3 \text{ and } 4 \leq j \leq 7. \tag{99}$$

Since Fz_1 and Fz_3 are orthogonal to M and Fz_2,Fz_4,Fz_5 are tangent to M, we can choose a $\mathcal{D} = S_p\{z_2, z_4, z_5\}$ and $\mathcal{D}^\perp = S_p\{z_1, z_3\}$. Thus, M is a 5−dimensional semi-invariant sub-manifold of $\mathbb{R}^7$ with usual almost Riemannian product structure $(F, <\,,\,>)$.

Example 3.2. Let M be sub-manifold of $\mathbb{R}^8$ by given

$$(u + v, u - v, u\cos\alpha, u\sin\alpha, u + v, u - v, u\cos\beta, u\sin\beta) \tag{100}$$

http://dx.doi.org/10.5772/65948

where u, v and β are the arbitrary parameters. By direct calculations, we can easily see that the tangent bundle of M is spanned by

$$\begin{aligned} e_1 &= \frac{\partial}{\partial x_1} + \frac{\partial}{\partial x_2} + \cos\alpha\frac{\partial}{\partial x_3} + \sin\alpha\frac{\partial}{\partial x_4} + \frac{\partial}{\partial x_5} - \frac{\partial}{\partial x_6} + \cos\beta\frac{\partial}{\partial x_7} + \sin\beta\frac{\partial}{\partial x_8} \\ e_2 &= \frac{\partial}{\partial x_1} - \frac{\partial}{\partial x_2} + \frac{\partial}{\partial x_5} + \frac{\partial}{\partial x_6}, e_3 = -u\sin\frac{\partial}{\partial x_3} + u\cos\alpha\frac{\partial}{\partial x_4}, \\ e_4 &= -u\sin\beta\frac{\partial}{\partial x_7} + u\cos\beta\frac{\partial}{\partial x_8}. \end{aligned} \tag{101}$$

For the almost Riemannian product structure F of $\mathbb{R}^8 = \mathbb{R}^4 x \mathbb{R}^4$, $F(TM)$ is spanned by vectors

$$\begin{aligned} Fe_1 &= \frac{\partial}{\partial x_1} + \frac{\partial}{\partial x_2} + \cos\alpha\frac{\partial}{\partial x_3} + \sin\alpha\frac{\partial}{\partial x_4} - \frac{\partial}{\partial x_5} + \frac{\partial}{\partial x_6} - \cos\beta\frac{\partial}{\partial x_7} - \sin\beta\frac{\partial}{\partial x_8}, \\ Fe_2 &= \frac{\partial}{\partial x_1} - \frac{\partial}{\partial x_2} - \frac{\partial}{\partial x_5} - \frac{\partial}{\partial x_6}, \ Fe_3 = e_3 \text{ and } Fe_4 = -e_4. \end{aligned} \tag{102}$$

Since Fe_1 and Fe_2 are orthogonal to M and Fe_3 and Fe_4 are tangent to M, we can choose $\mathcal{D}^T = Sp\{e_3, e_4\}$ and $\mathcal{D}^\perp = Sp\{e_1, e_2\}$. Thus, M is a four-dimensional semi-invariant sub-manifold of $\mathbb{R}^8 = \mathbb{R}^4 x \mathbb{R}^4$ with usual Riemannian product structure F.

Definition 3.3. Let M be a sub-manifold of an almost Riemannian product manifold $\tilde{M}$ with almost Riemannian product structure $(F, \tilde{g})$. M is said to be pseudo-slant sub-manifold if there exist distributions $\mathcal{D}_\theta$ and $\mathcal{D}_\perp$ on M such that

i. The tangent bundle $TM = \mathcal{D}_\theta \oplus \mathcal{D}^\perp$.

ii. The distribution $\mathcal{D}_\theta$ is a slant distribution with slant angle θ.

iii. The distribution $\mathcal{D}^\perp$ is an anti-invariant distribution, i.e., $F(\mathcal{D}^\perp) \subseteq T^\perp M$.

As a special case, if $\theta = 0$ and $\theta = \frac{\pi}{2}$, then pseudo-slant sub-manifold becomes semi-invariant and anti-invariant sub-manifolds, respectively.

Example 3.3. Let M be a sub-manifold of $\mathbb{R}^6$ by the given equation

$$(\sqrt{3}u, v, v\sin\theta, v\cos\theta, s\cos t, -s\cos t) \tag{103}$$

where u, v, s and t arbitrary parameters and θ is a constant.

We can check that the tangent bundle of M is spanned by the tangent vectors

$$\begin{aligned} e_1 &= \sqrt{3}\frac{\partial}{\partial x_1}, e_2 = \frac{\partial}{\partial y_1} + \sin\theta\frac{\partial}{\partial x_2} + \cos\theta\frac{\partial}{\partial y_2}, \\ e_3 &= \cos t\frac{\partial}{\partial x_3} - \cos t\frac{\partial}{\partial y_3}, e_4 = -s\sin t\frac{\partial}{\partial x_3} + s\sin t\frac{\partial}{\partial y_3}. \end{aligned} \tag{104}$$

For the almost product Riemannian structure F of $\mathbb{R}^6$ whose coordinate systems $(x_1, y_1, x_2, y_2, x_3, y_3)$ choosing

$$F\left(\frac{\partial}{\partial x_i}\right) = \frac{\partial}{\partial y_i}, 1 \le i \le 3,$$
$$F\left(\frac{\partial}{\partial y_j}\right) = \frac{\partial}{\partial x_j}, 1 \le j \le 3, \tag{105}$$

Then, we have

$$Fe_1 = \sqrt{3}\frac{\partial}{\partial y_1}, Fe_2 = -\frac{\partial}{\partial x_1} + \sin\theta\frac{\partial}{\partial y_2} - \cos\theta\frac{\partial}{\partial x_2}$$
$$Fe_3 = \cos t\frac{\partial}{\partial y_3} + \cos t\frac{\partial}{\partial x_3}, Fe_4 = -s\sin t\frac{\partial}{\partial y_3} - s\sin t\frac{\partial}{\partial x_3}. \tag{106}$$

Thus, $\mathcal{D}_\theta = S_p\{e_1, e_2\}$ is a slant distribution with slant angle $\alpha = \frac{\pi}{4}$. Since Fe_3 and Fe_4 are orthogonal to M, $\mathcal{D}^\perp = S_p\{e_3, e_4\}$ is an anti-invariant distribution, that is, M is a 4-dimensional proper pseudo-slant sub-manifold of $\mathbb{R}^6$ with its almost Riemannian product structure $(F, <, >)$.

Author details

Mehmet Atçeken[1*], Ümit Yıldırım[1] and Süleyman Dirik[2]

*Address all correspondence to: mehmet.atceken382@gmail.com

1 Department of Mathematics, Faculty of Arts and Sciences, Gaziosmanpasa University, Tokat, Turkey

2 Department of Statistic, Faculty of Arts and Sciences, Amasya University, Amasya, Turkey

References

[1] Katsuei Kenmotsu, editor. Differential Geometry of Submaifolds. Berlin: Springe-Verlag; 1984. 134 p.

[2] Aurel Bejancu. Geometry of CR-Submanifolds. Dordrecht: D. Reidel Publishing Company; 1986. 172 p. DOI: QA649.B44

[3] Bang-Yen Chen. Geometry of Submanifolds. New York: Marcel Dekker, Inc.; 1973. 298 p.

[4] Kentaro Yano and Masahiro Kon. Structurs on Manifolds. Singapore: World Scientific Publishing Co. Pte. Ltd.; 1984. 508 p. DOI: QA649.Y327

[5] Meraj Ali Khan. Geometry of Bi-slant submanifolds "Some geometric aspects on submanifolds Theory". Saarbrücken, Germany: Lambert Academic Publishing; 2006. 112 p.

[6] Mehmet Atçeken. Warped product semi-invariant submanifolds in almost paracontact Riemannian manifolds. Mathematical Problems in Engineering. 2009;2009:621625. DOI: doi:10.1155/2009/621625

[7] Tyuzi Adati. Submanifolds of an almost product Riemannian manifold. Kodai Mathematical Journal. 1981;4(2):327–343.

[8] Mehmet Atçeken. A condition for warped product semi-invariant submanifolds to be Riemannian product semi-invariant Sub-manifoldsub-manifolds in locally Riemannian product manifolds. Turkish Journal of Mathematics. 2008;**33**:349–362.

[9] Mehmet Atçeken. Slant submanifolds of a Riemannian product manifold. Acta Mathematica Scientia. 2010;**30**(1):215–224. DOI: doi:10.1016/S0252-9602(10)60039-2

Chapter 4

Spectral Theory of Operators on Manifolds

Paul Bracken

Additional information is available at the end of the chapter

http://dx.doi.org/10.5772/64611

Abstract

Differential operators that are defined on a differentiable manifold can be used to study various properties of manifolds. The spectrum and eigenfunctions play a very significant role in this process. The objective of this chapter is to develop the heat equation method and to describe how it can be used to prove the Hodge Theorem. The Minakshisundaram-Pleijel parametrix and asymptotic expansion are then derived. The heat equation asymptotics can be used to give a development of the Gauss-Bonnet theorem for two-dimensional manifolds.

Keywords: manifold, operator, differential form, Hodge theory, eigenvalue, partial differential operator, Gauss-Bonnet

1. Introduction

Topological and geometric properties of a manifold can be characterized and further studied by means of differential operators, which can be introduced on the manifold. The only natural differential operator on a manifold is the exterior derivative operator which takes k-forms to $k+1$ forms. This operation is defined purely in terms of the smooth structure of the manifold, used to define de Rham cohomology groups. These groups can be related to other topological quantities such as the Euler characteristic. When a Riemannian metric is defined on the manifold, a set of differential operators can be introduced. The Laplacian on k-forms is perhaps the most well known, as well as other elliptic operators.

On a compact manifold, the spectrum of the Laplacian on k-forms contains topological as well as geometric information about the manifold. The Hodge theorem relates the dimension of the kernel of the Laplacian to the k-th Betti number requiring them to be equal. The Laplacian determines the Euler characteristic of the manifold. A sophisticated approach to obtaining information related to the manifold is to consider the heat equation on k-forms with its solution given by the heat semigroup [1–3].

The heat kernel is one of the more important objects in such diverse areas as global analysis, spectral geometry, differential geometry, as well as in mathematical physics in general. As an example from physics, the main objects that are investigated in quantum field theory are described by Green functions of self-adjoint, elliptic partial differential operators on manifolds as well as their spectral invariants, such as functional determinants. In spectral geometry, there is interest in the relation of the spectrum of natural elliptic partial differential operators with respect to the geometry of the manifold [4–6].

Currently, there is great interest in the study of nontrivial links between the spectral invariants and nonlinear, completely integrable evolutionary systems, such as the Korteweg-de Vries hierarchy. In many interesting situations, these systems are actually infinite-dimensional Hamiltonian systems. The spectral invariants of a linear elliptic partial differential operator are nothing but the integrals of motion of the system. There are many other applications to physics such as to gauge theories and gravity [7].

In general, the existence of nonisometric isospectral manifolds implies that the spectrum alone does not determine the geometry entirely. It is also important to study more general invariants of partial differential operators that are not spectral invariants. This means that they depend not only on the eigenvalues but also on the eigenfunctions of the operator. Therefore, they contain much more information with respect to the underlying geometry of the manifold.

The spectrum of a differential operator is not only studied directly, but the related spectral functions such as the spectral traces of functions of the operator, such as the zeta function and the heat trace, are relevant as well [8, 9]. Often the spectrum is not known exactly, which is why different asymptotic regimes are investigated [10, 11]. The small parameter asymptotic expansion of the heat trace yields information concerning the asymptotic properties of the spectrum. The trace of the heat semigroup as the parameter approaches zero is controlled by an infinite sequence of geometric quantities, such as the volume of the manifold and the integral of the scalar curvature of the manifold. The large parameter behavior of the traces of the heat kernels is parameter independent and in fact equals the Euler characteristic of the manifold. The small parameter behavior is given by an integral of a complicated curvature-dependent expression. It is quite remarkable that when the dimension of the manifold equals two, the equality of the short- and long-term behaviors of the heat flow implies the classic Gauss-Bonnet theorem. The main objectives of the chapter are to develop the heat equation approach with Schrödinger operator on a vector bundle and outline how it leads to the Hodge theorem [12, 13]. The heat equation asymptotics will be developed [14, 15] andit is seen that the Gauss-Bonnet theorem can be proved for a two-dimensional manifold based on it. Moreover, this kind of approach implies that there is a generalization of the Gauss-Bonnet theorem as well in higher dimensions greater than two [16, 17].

2. Geometrical preliminaries

For an n-dimensional Riemannian manifold M, an orthonormal moving frame $\{e_1,\dots,e_n\}$ can be chosen with $\{\omega_1,\dots,\omega_n\}$ the accompanying dual coframe which satisfy

http://dx.doi.org/10.5772/64611

$$\omega_i(e_j) = \delta_{ij}, \qquad i,j = 1,\ldots,n \tag{1}$$

It is then possible to define a system of one-forms ω_{ij} and two-forms Ω_{ij} by solving the equations,

$$\nabla_X e_i = \sum_j \omega_{ji}(X)\, e_j, \qquad R(X,Y)e_i = \sum_j \Omega_{ji}(X,Y)e_j \tag{2}$$

It then follows that the Christoffel coefficients and components of the Riemann tensor for M are

$$\omega_{ji}(e_k) = \sum_a \langle \omega_{aj}(e_k)e_a, e_i\rangle_g = \langle \nabla_{e_k} e_j, e_i\rangle_g = \Gamma^i_{kj} \tag{3}$$

$$\Omega_{ij}(e_k,e_s) = \sum_a \langle \Omega_{aj}(e_k,e_s)e_a, e_i\rangle_g = \langle\, R(e_k,e_s)e_j, e_i\rangle_g = R_{ksji} \tag{4}$$

The inner product induced by the Riemannian metric on M is denoted here by $\langle\cdot,\cdot\rangle : \Gamma(TM) \times\Gamma(TM) \to \mathcal{F}(M)$ and it induces a metric on $\Lambda^k(M)$ as well. Using the Riemannian metric and the measure on M, an inner product denoted $\langle\langle\cdot,\cdot\rangle\rangle : \Lambda^k(M)\times\Lambda^k(M) \to \mathbb{R}$ can be defined on $\Lambda^k(M)$ so that for $\alpha,\beta\in\Lambda^k(M)$,

$$\langle\langle\alpha,\beta\rangle\rangle = \int_M \langle\alpha,\beta\rangle_g \, dv_M \tag{5}$$

where if $(x^1,\ldots,x^m)$ is a system of local coordinates,

$$dv_M = \det(g_{ij})\, dx^1\wedge\ldots\wedge dx^m$$

is the Riemannian measure on M. Clearly, $\langle\langle\alpha,\beta\rangle\rangle$ is linear with respect to α, β and $\langle\langle\alpha,\alpha\rangle\rangle\geq 0$ with equality if and only if $\alpha = 0$. Hodge introduced a star homomorphism $* : \Lambda^k(M) \to \Lambda^{n-k}(M)$, which is defined next.

Definition 2.1. (i) For $\omega = \sum_{i_1<\cdots<i_k} f_{i_1\cdots i_k}\, \omega_{i_1}\wedge\cdots\omega_{i_k}$, define

$$^*\omega = \sum_{\substack{i_1<\cdots<i_k \\ j_1<\cdots<j_{n-k}}} f_{i_1\cdots i_k}\epsilon(i_1,\ldots,i_k,j_1,\ldots,j_{n-k})\omega_{j_1}\wedge\cdots\wedge\omega_{j_{n-k}},$$

where ϵ is 1, −1, or 0 depending on whether $(i_1,\ldots,i_k,j_1,\ldots,j_{n-k})$ is an even or odd permutation of $(1,\ldots,n)$, respectively.

(ii) If M is an oriented Riemannian manifold with dimension n, define the operator

$$\delta = (-1)^{nk+n+1*}d^* : \Lambda^k(M) \to \Lambda^{k-1}(M) \tag{6}$$

In terms of the two operators d and δ, the Laplacian acting on k-forms can be defined on the two subspaces

$$\Lambda^{\text{even}}(M) = \oplus_{\text{even}} \Lambda^k(M), \qquad \Lambda^{\text{odd}}(M) = \oplus_{\text{odd}} \Lambda^k(M) \tag{7}$$

The operator $d + \delta$ can be regarded as the operators on these subspaces,

$$D_0 = d + \delta : \Lambda^{\text{even}}(M) \to \Lambda^{\text{odd}}(M), \quad D_1 = d + \delta : \Lambda^{\text{odd}}(M) \to \Lambda^{\text{even}}(M) \tag{8}$$

Definition 2.2. Let M be a Riemannian manifold, then the operator

$$D_0 = d + \delta : \Lambda^{\text{even}}(M) \to \Lambda^{\text{odd}}(M) \tag{9}$$

is called the Hodge-de Rham operator. It has the property that it is a self-conjugate operator, $D_0^* = D_1$ and $D_1^* = D_0$. It is useful in studying the Laplacian to have a formula for the operator $\Delta = (d + \delta)^2$ and hence for $D_0^* D_0$ and $D_1^* D_1$ as well.

Let $\{e_1,\ldots,e_n\}$ be an orthonormal moving frame defined on an open set U. Define as well the pair of operators

$$E_j^+ = \omega_j \wedge \cdot + i(e_j) : \Lambda^*(U) \to \Lambda^*(U), \qquad E_j^- = \omega_j \wedge \cdot - i(e_j) : \Lambda^*(U) \to \Lambda^*(U) \tag{10}$$

Lemma 2.1. The operators $E_j^{\pm}$ satisfy the following relations

$$E_i^+ E_j^+ + E_j^+ E_i^+ = 2\delta_{ij}, \quad E_i^+ E_j^- + E_j^- E_i^+ = 0, \quad E_i^- E_j^- + E_j^- E_i^- = -2\delta_{ij} \tag{11}$$

If M is a Riemannian manifold and $\nabla : \Gamma(TM) \times \Gamma(TM) \to \Gamma(TM)$ is a Levi-Civita connection, then a connection on the space $\Lambda^*(M)$, namely $(X,\omega) \to \nabla_X \omega$, can also be defined such that

$$(\nabla_X \omega)(Y) = X(\omega(Y)) - \omega(\nabla_X Y), \quad Y \in \Gamma(TM)$$

The connection may be regarded as a first-order derivative operator $(X,Y,\omega) \to D(X,Y)\omega$.

Definition 2.3. The second-order derivative operator $(X,Y,\omega) \to D(X,Y)\omega$ is defined to be

$$D(X,Y)\omega = \nabla_X \nabla_Y \omega - \nabla_{\nabla_X Y} \omega \tag{12}$$

In terms of the operator (Eq. (12)), define a second-order differential operator $\Delta_0 : \Lambda^*(M) \to \Lambda^*(M)$ by

$$\Delta_0 = \sum_i D(e_i, e_i), \tag{13}$$

where $\{e_i\}_1^n$ is an orthonormal moving frame. The operator Δ_0 in Eq. (13) is referred to as the *Laplace-Beltrami* operator.

Theorem 2.1. (Weitzenböck) Let M be a Riemannian manifold M with an associated orthonormal moving frame $\{e_i\}_1^n$. The Laplace operator can be expressed as

$$\Delta = (d+\delta)^2 = -\Delta_0 - \frac{1}{8} \sum_{i,j,k,s} R_{ijks}\, E_i^+ E_j^+ E_k^- E_s^- + \frac{1}{4} R \tag{14}$$

In Eq. (14), R is the scalar curvature, $R = -\sum_{i,j} R_{ijij}$ and Δ_0 is the Laplace-Beltrami operator (13).

The operator defined by Eq. (14) does not contain first-order covariant derivatives and is of a type called a Schrödinger operator. Thus, Weitzenböck formula (14) implies the that Laplacian can be expressed in the form $\Delta = -\Delta_0 - F$ and is an elliptic operator. The Schrödinger operator (14) can be used to define an operator that plays an important role in mathematical physics. The heat operator is defined to be

$$\mathcal{H} = \frac{\partial}{\partial t} + \Delta \tag{15}$$

The crucial point for the theory of the heat operator is the existence of a fundamental solution. In fact, the Hodge theorem can be proved by making use of the fundamental solution.

Definition 2.4. Let M be a Riemannian manifold, $\pi : E \to M$ is a vector bundle with connection. Let $\Delta_0 : \Gamma(E) \to \Gamma(E)$ be the Laplace-Beltrami operator, which is defined by means of the Levi-Civita connection on M and the connection on the vector bundle E. Let $F : \Gamma(E) \to \Gamma(E)$ be a $\mathcal{F}(M)$-linear map. Then, $\Delta = -\Delta_0 - F$ is a Schrödinger operator. If a family of R-linear maps

$$G(t,q,p) : E_p \to E_q$$

with parameter $t > 0$ and $q,p \in M$ satisfies the following three conditions, the family is called a fundamental solution of the heat operator (15) where $E_p = \pi^{-1}(p)$. First, $G(t,q,p) : E_p \to E_q$ is an R-linear map of vector spaces and continuous in all variables t,q,p. Second, for a fixed $w \in E_p$, let $\theta(t,q) = G(t,q,p)w$, for all $t > 0$, then θ has first and second continuous derivatives in t and q, respectively andsatisfies the heat equation, which for $t > 0$ is given by $\mathcal{H}\theta(t,q) = 0$, which can be written as

$$\left(\frac{\partial}{\partial t} + \Delta_q\right) G(t,q,p) = 0 \tag{16}$$

where Δ_q acts on the variable q. Finally, if φ is a continuous section of the vector bundle E, then

$$\lim_{t \to 0^+} \int_M G(t,q,p)\varphi(p)\, dv_p = \varphi(q)$$

for all φ, where dv_p is the volume measure with respect to the coordinates of p given in terms of the Riemannian metric.

Definition 2.5. Suppose a $G_0(t,q,p)$ is given. The following procedure taking $G_0(t,q,p)$ to $G(t,q,p)$ is called the Levi algorithm:

$$K_0(t,q,p) = \left(\frac{\partial}{\partial t} + \Delta_q\right) G(t,q,p),$$

$$K_{m+1}(t,q,p) = \int_0^t d\tau \int_M K_0(t-\tau,q,z) K_m(\tau,z,p)\, dv_z$$

$$\overline{K}(t,q,p) = \sum_{m=0}^{\infty} (-1)^{m+1} K_m(t,q,p), \tag{17}$$

$$G(t,q,p) = G_0(t,q,p) + \int_0^t d\tau \int_M G_0(t-\tau,q,z) \overline{K}(\tau,z,p)\, dv_z$$

The Cauchy problem can be formulated for the heat equation such that existence, regularity and uniqueness of solution can be established. The Hilbert-Schmidt theorem can be invoked to develop a Fourier expansion theorem applicable to this Schrödinger operator.

Suppose $\Delta : \Gamma(E) \rightarrow \Gamma(E)$ is a self-adjoint nonnegative Schrödinger operator, then there exists a set of C^∞ sections $\{\psi_i\} \subset \Gamma(E)$ such that

$$\langle\langle \psi_i, \psi_j \rangle\rangle = \int_M \langle \psi_i(x), \psi_j(x) \rangle\, dv_x = \delta_{ij}$$

Moreover, denoting the completion of the inner product space $\Gamma(E)$ by $\overline{\Gamma(E)}$, the set $\{\psi_i\}$ is a complete set in $\overline{\Gamma(E)}$, so for any $\psi \in \overline{\Gamma(E)}$,

$$\psi = \sum_{i=1}^{\infty} \langle\langle \psi, \psi_i \rangle\rangle\, \psi_i$$

Finally, the set $\{\psi_i\}$ satisfies the equation

$$\Delta \psi_i = \lambda_i \psi_i, \qquad T_t \psi_i = e^{-t\lambda_i} \psi_i$$

where λ_i are the eigenvalues of Δ andform an increasing sequence: $0 \leq \lambda_1 \leq \lambda_2 \leq \cdots$ where $\lim_{k\rightarrow\infty} \lambda_k = \infty$.

Denote $U(t,q)$ by $(T_t\psi)(q)$ when $U(0,q) = \psi(q)$ and T_t satisfies the semigroup property andT_t is a self-adjoint, compact operator.

Theorem 2.2. Let $G(t,q,p)$ be the fundamental solution of the heat operator (15), then

$$G(t,q,p)w = \sum_{i=1}^{\infty} e^{\lambda_i t} \langle\, \psi_i(p), w \rangle \psi_i(q) \tag{18}$$

with $w \in E_p$ holds in $\overline{\Gamma(E)}$.

Proof: For fixed $t > 0$ and $w \in E_p$, expand $G(t,q,p)w$ in terms of eigenfunctions $\psi_i(q)$,

$$G(t,q,p)w = \sum_{i=1}^{\infty} \sigma_i(t,p,w)\psi_i(q), \quad \sigma_i(t,p,w) = \int_M \langle \psi_i(q), G(t,q,p)w \rangle \, dv_q$$

Differentiating with respect to t and using $\Delta\psi_i = \lambda_i\psi_i$, we get

$$\begin{aligned} \frac{\partial}{\partial t}\sigma_i(t,p,w) &= \int_M \langle \psi_i(q), \frac{\partial}{\partial t}G(t,q,p)w \rangle \, dv_q = \int_M \langle \psi_i(q), -\Delta_q G(t,q,p)w \rangle dv_q \\ &= -\int_M \langle \Delta_q \psi_i(q), G(t,q,p)w \rangle \, dv_q = -\lambda_i \int_M \langle \psi_i(q), G(t,q,p)w \rangle \, dv_q \\ &= -\lambda_i \sigma_i(t,p,w) \end{aligned}$$

It follows from this that

$$\sigma_i(t,p,w) = c_i(p,w)e^{-\lambda_i t}$$

and since σ_i depend linearly on w, so $c_i(p,w) = c_i(p)w$, where $c_i(p) : E_p \to \mathbb{R}$ is a linear function. There exists $\tilde{c}_i(p)$ independent of w such that $c_i(p)w = \langle \tilde{c}_i(p), w \rangle$ so that

$$G(t,q,p)w = \sum_{i=1}^{\infty} e^{\lambda_i t} \, \psi_i(q) \langle \tilde{c}_i(p), w \rangle$$

Consequently, for any $\beta \in \Gamma(E)$, we have

$$\beta(q) = \lim_{t \to 0} \int_M G(t,q,p)\beta(p) \, dv_p = \sum_{k=1}^{\infty} \psi_k(q) \int_M \langle \tilde{c}_k(p), \beta(p) \rangle \, dv_p$$

Moreover, $\beta(q)$ can also be expanded in terms of the ψ_k basis set,

$$\beta(q) = \sum_{k=1}^{\infty} \psi_k(q) \int_M \langle \psi_k(p), \beta(p) \rangle \, dv_p$$

Upon comparing these last two expressions, it is clear that $\tilde{c}_k(p) = \psi_k(p)$ for all k andwe are done.

One application of the heat equation method developed so far is to develop and give a proof of the Hodge theorem.

Theorem 2.3. Let M, E, Δ be defined as done already, then

1. $H = \{\varphi \in \Gamma(E) | \Delta\varphi = 0\}$ is a finite-dimensional vector space.
2. For any $\psi \in \Gamma(E)$, there is a unique decomposition of ψ as $\psi = \psi_1 \oplus \psi_2$, where $\psi_1 \in H$ and $\psi_2 \in \Delta(\Gamma(E))$.

The first part is a direct consequence of the expansion theorem and due to the fact $H \perp \Delta(\Gamma(E))$, the decomposition is unique.

The Hodge theorem has many applications, but one in particular fits here. It is used in conjunction with the de Rham cohomology group $H^*_{dR}(M)$. Define

$$Z^k(M) = \ker\{d : \Lambda^k(M) \to \Lambda^{k+1}(M)\} \equiv \{\alpha \in \Lambda^k(M) | \, d\alpha = 0\} \tag{19}$$

$$B^k(M) = \ \mathrm{Im}\,\{d : \Lambda^{k-1}(M) \to \Lambda^k(M)\} \equiv d(\Lambda^{k-1}(M)) \tag{20}$$

Since $d^2 = 0$, it follows that $B^k(M) \subset Z^k(M)$ andthe k-th de Rham cohomology group of M is defined to be

$$H^k_{dR}(M) = Z^k(M)/B^k(M) \tag{21}$$

From Eq. (21), construct

$$H^*_{dR}(M) = \oplus_k \, H^k_{dR}(M) \tag{22}$$

In 1935, Hodge claimed a theorem, which stated every element in $H^k_{dR}(M)$ can be represented by a unique harmonic form α, one which satisfies both $d\alpha = 0$ and $\delta\alpha = 0$. Denote the set of harmonic forms as $H^k(M)$.

Theorem 2.4. Let M be a Riemannian manifold of dimension n, then

$$H^k(M) = \ker\,\{d + \delta : \Lambda^k(M) \to \Lambda^*(M)\} = \ker\,\{\Delta : \Lambda^k(M) \to \Lambda^k(M)\} \tag{23}$$

where $\Delta = (d + \delta)^2$.

Proof: Since $\Delta = d\delta + \delta d$, this implies that $\Delta(\Lambda^k(M)) \subset \Lambda^k(M)$ andit is clear that

$H^k(M) \subset \ker\{d + \delta : \Lambda^k(M) \to \Lambda^*(M)\} \subset \ker\{\Delta : \Lambda^k(M) \to \Lambda^*(M)\} = \ker\,\{\Delta : \Lambda^k(M) \to \Lambda^k(M)\}$. To finish the proof, it suffices to show that $\ker\{\Delta : \Lambda^k(M) \to \Lambda^k(M)\} \subset H^k(M)$. If $\alpha \in \ker\{\Delta : \Lambda^k(M) \to \Lambda^k(M)\}$, that is $\Delta\alpha = 0$, then

$$\begin{aligned}\langle\langle \Delta\alpha,\alpha,\rangle\rangle = \langle\langle (d+\delta)^2\alpha,\alpha\rangle\rangle = \langle\langle (d+\delta)\alpha,(d+\delta)\alpha\rangle\rangle &= \langle\langle d\alpha,d\alpha\rangle\rangle + \langle\langle \delta\alpha,\delta\alpha\rangle\rangle + 2\langle\langle d\alpha,\delta\alpha\rangle\rangle \\ &= \langle\langle d\alpha,d\alpha\rangle\rangle + \langle\langle \delta\alpha,\delta\alpha\rangle\rangle = 0\end{aligned}$$

This implies that $d\alpha = 0$ and $\delta\alpha = 0$, hence $\alpha \in H^k(M)$.

Theorem 2.5. Let M be a Riemannian manifold of dimension n, then

1. $H^k(M)$ is a finite dimensional vector space for $k = 0, 1, 2, \ldots, n$.
2. There is an orthogonal decomposition of $\Lambda^k(M)$ as

$$\Lambda^k(M) = H^k(M) + d(\Lambda^{k-1}(M)) + \delta(\Lambda^{k+1}(M)) \tag{24}$$

http://dx.doi.org/10.5772/64611

Proof: By Theorem 2.1, $\Delta : \Lambda^k(M) \to \Lambda^k(M)$ is a Schrödinger operator, so the Hodge theorem applies. Thus $H^k(M)$ is of finite dimension, so the first holds. The second part of the Hodge theorem is $\Lambda^k(M) = H^k(M) + \Delta(\Lambda^k(M))$. Since $\Delta(\Lambda^k(M)) \subset d(\Lambda^{k-1}(M)) + \delta(\Lambda^{k+1}(M))$, we have $\Lambda^k(M) = H^k(M) + d(\Lambda^{k-1}(M)) + \delta(\Lambda^{k+1}(M))$. The three spaces in this decomposition are orthogonal to each other, so (ii) holds as well.

Theorem 2.6. (Duality theorem) For an oriented Riemannian manifold M of dimension n, the star isomorphism $* : H^k(M) \to H^{n-k}(M)$ induces an isomorphism

$$H^k_{dR}(M) \simeq H^{n-k}_{dR}(M) \tag{25}$$

The k-th Betti number defined as $b_k(M) = \dim H^k(M, \mathbb{R})$ also satisfies $b_k(M) = b_{n-k}(M)$ for $0 \leq k \leq n$.

3. The Minakshisundaran-Pleijel paramatrix

Let M be a Riemannian manifold with dimension n and E a vector bundle over M with an inner product and a metric connection. Here, the following formal power series is considered with a special transcendental multiplier $e^{-\rho^2/4t}$ and parameters $(t,p,q) \in (0,\infty) \times M \times M$, defined by

$$H_\infty(t,q,p) = \frac{1}{(4\pi t)^{n/2}} e^{-\rho^2/4t} \sum_{k=0}^{\infty} t^k\, u_k(p,q) : E_p \to E_q \tag{26}$$

In Eq. (26), the function $\rho = \rho(p,q)$ is the metric distance between p and q in M, $E_p = \pi^{-1}(p)$ is the fiber of E over p and $u_k(p,q) : E_p \to E_q$ are R-linear map.

It is the objective to find conditions for which Eq. (26) satisfies the heat equation or the following equality:

$$\left(\frac{\partial}{\partial t} + \Delta_q\right) H_\infty(t,q,p) w = 0 \tag{27}$$

To carry out this, a normal coordinate system denoted by $\{x_1, \ldots, x_n\}$ is chosen in a neighborhood of point p and is centered at p. This means that if q is in this neighborhood about p, which has coordinates $(x_1, \ldots, x_n)$, then the function $\rho(p,q)$ is

$$\rho(p, q) = \sqrt{x_1^2 + \cdots + x_n^2} \tag{28}$$

In terms of these coordinates, we calculate the components of g,

$$g_{ij} = \left\langle \frac{\partial}{\partial x_i}, \frac{\partial}{\partial x_j} \right\rangle, \qquad G = \det(g_{ij}) \tag{29}$$

and define the differential operator

$$\hat{\partial} = \sum_{k=1}^{n} x_k \frac{\partial}{\partial x_k}$$

The notion of the heat operator (15) on Eq. (26) is worked out one term at a time. First, the derivative with respect to t is calculated

$$\begin{aligned} \frac{\partial}{\partial t} H_\infty(t,p,q)w &= \frac{1}{(4\pi t)^{n/2}} e^{-\rho^2/4t} \left\{ \left(\frac{\rho^2}{4t^2} - \frac{n}{2t} \right) \sum_{k=0}^{\infty} t^k u_k(p,q)w + \sum_{k=0}^{\infty} k t^{k-1} u_k(p,q)w \right\} \\ &= \frac{1}{(4\pi t)^{n/2}} e^{-\rho^2/4t} \sum_{k=0}^{\infty} \left\{ \frac{\rho^2}{4t^2} - \frac{n}{2t} + \frac{k}{t} \right\} t^k u_k(p,q)w \end{aligned} \tag{30}$$

It is very convenient to abbreviate the function appearing in front of the sum in Eq. (30) as follows:

$$\Phi(\rho) = \frac{e^{-\rho^2/4t}}{(4\pi t)^{n/2}} \tag{31}$$

Let $\{e_1,\ldots,e_n\}$ be a frame that is parallel along geodesics passing through p and satisfies

$$e_i(p) = \frac{\partial}{\partial x_i}\Big|_p$$

In terms of the function in Eq. (31), the operator Δ_0 acting on Eq. (26) is given as

$$\begin{aligned} \Delta_0 H_\infty(t,p,q)w &= (\Delta_0\, \Phi) \cdot \left(\sum_{k=0}^{\infty} t^k u_k(p,q)w \right) \\ &+2 \sum_{a=1}^{n} (e_a\Phi) \cdot \nabla_{e_a} \left(\sum_{k=0}^{\infty} t^k u_k(p,q)w \right) + \Phi \cdot \Delta_0 \left(\sum_{k=0}^{\infty} t^k u_k(p,q)w \right) \end{aligned} \tag{32}$$

The individual components of (32) can be calculated as follows; since Φ is a function $\nabla_{e_a} \Phi = e_a \Phi$ and so

$$\begin{aligned} e_a\Phi(\rho) &= \Phi'(\rho) e_a(\rho), \\ \Delta_0 \Phi &= \sum_a \{ e_a\, e_a \Phi(\rho) - (\nabla_{e_a} e_a)\Phi(\rho) \} = \Phi''(\rho) \cdot \sum_a (e_a\rho)^2 + \Phi'(\rho) \cdot \Delta_0 \rho, \\ \Phi'(\rho) &= -\frac{\rho}{2t} \Phi(\rho), \\ \Phi''(\rho) &= \left(\frac{\rho^2}{4t^2} - \frac{1}{2t} \right) \Phi(\rho) \end{aligned} \tag{33}$$

Consequently,

$$e_a\rho = \frac{x_a}{\rho}, \qquad \sum_a (e_a\rho)^2 = 1, \qquad \Delta_0\rho = \frac{n-1}{\rho} + \frac{1}{\rho}\hat{\partial}\ \log\sqrt{G}$$

and the Laplace-Beltrami operator on the function Φ is given by

$$\Delta_0\,\Phi = \Phi(\rho)\left(\left(\frac{\rho^2}{4t^2} - \frac{1}{2t}\right) - \frac{1}{2t}(n-1-\hat{\partial}\log\sqrt{G})\right) \tag{34}$$

Expression (34) goes into the first term on the right side of Eq. (32). The second term on the right-hand side of (32) takes the form,

$$\begin{aligned} 2\sum_{a=1}^{n}(e_a\Phi)\cdot\nabla_{e_a}\left(\sum_{k=0}^{\infty} t^k u_k(p,q)w\right) &= 2\Phi'(\rho)\sum_{a=1}^{n}\frac{x_a}{\rho}\cdot\nabla_{e_a}\left(\sum_{k=0}^{\infty} t^k u_k(p,q)w\right) \\ &= -\frac{\rho}{t}\Phi(\rho)\nabla_{\hat{\partial}/\rho}\left(\sum_{k=0}^{\infty} t^k u_k(p,q)w\right) \end{aligned} \tag{35}$$

Substituting these results into (32), it follows that

$$\Delta_0\,H_\infty(t,q,p) = \Phi(\rho)\left[\frac{\rho^2}{4t^2} - \frac{1}{2t} - \frac{1}{2t}(n-1-\hat{\partial}\log\sqrt{G}) - \frac{\rho}{t}\nabla_{\hat{\partial}/\rho} + \Delta_0\right]\sum_{m=0}^{\infty} t^m u_m(p,q)w \tag{36}$$

Combining Eq. (36) with the derivative of H_∞ with respect to t in Eq. (35), the following version of the heat equation results:

$$\begin{aligned} \left(\frac{\partial}{\partial t} - \Delta_0 - F\right)H_\infty(t,q,p)w = \Phi\Bigg[&\left(\nabla_{\hat{\partial}} + \frac{1}{4G}\hat{\partial}G\right)\cdot\frac{1}{t}u_0(p,q)w + \sum_{k=1}^{\infty}\Bigg[\left(\nabla_{\hat{\partial}} + k + \frac{1}{4G}\hat{\partial}G\right)u_k(p,q)w \\ &-(\Delta_0 + F)u_{k-1}(p,q)w\Bigg]t^{k-1}\Bigg] \end{aligned} \tag{37}$$

This is summarized in the following Lemma.

Lemma 3.1. Heat equation (27) for $H_\infty(t,p,q)$ is equivalent to

$$\left(\nabla_{\hat{\partial}} + k + \frac{1}{4G}\hat{\partial}G\right)u_k(p,q)w = (\Delta_0 + F)\,u_{k-1}(p,q)w \tag{38}$$

for all $k = 0, 1, 2, \ldots$ and Eq. (38) is initialized with $u_{-1}(p,q) = 0$.

In fact, for fixed $p \in M$ and $w \in E_p$, there always exists a unique solution to problem (Eq. (38)) over a small coordinate neighborhood about p.

Definition 3.1. Denote the solution of Eq. (38) by $u(p,q)w$, which depends linearly on w. Then, $u_m(p,q) : E_p \to E_q$ and the Minakshisundaram-Pleijel parametrix for heat operator (Eq. 15) is defined by

$$H_\infty(t,p,q) = \frac{1}{(4\pi t)^{n/2}} e^{-\rho^2/4t} \sum_{m=0}^{\infty} t^m u_m(p,q) : E_p \to E_q \tag{39}$$

Based on Eq. (39), the N-truncated parametrix is defined based on Eq. (39) to be

$$H_N(t,q,p) = \frac{1}{(4\pi t)^{n/2}} e^{-\rho^2/4t} \sum_{m=0}^{N} t^m u_m(p,q) : E_p \to E_q \tag{40}$$

Theorem 3.1. Choose a smooth function $\phi : M\times M \to M$ and let $G_0(t,q,p) = \phi(q,p)H_N(t,q,p)$. Then $G_0(t,q,p)$ is a k-th initial solution of the heat operator (15), where $k = \lfloor \frac{N}{2} - \frac{n}{4} \rfloor$ and $\lfloor z \rfloor$ is the greatest integer less than or equal to z.

Proof: Clearly, G_0 is a linear map of vector spaces andis continuous and C^∞ in all parameters. From the previous calculation, it holds that

$$\left(\frac{\partial}{\partial t} - \Delta_0 - F\right) H_N(t,q,p)w = -\frac{1}{(4\pi t)^{n/2}} e^{-\rho^2/4t} t^{N-\frac{n}{2}} (\Delta_0 + F) u_N(p,q)w \tag{41}$$

and $u_N(p,q)$ is C^∞ with respect to p and q. Since $t^{N-\frac{n}{2}} e^{-\rho^2/4t}$ is $C^k([0,\infty)\times M\times M)$, hence $\mathcal{H}(\varphi(p,q)H_N(t,q,p)) \in C^k([0,\infty)\times M\times M)$. Consider integrating G_0 against $\psi(s,\beta)$,

$$\int_M G_0(t,q,s)\psi(s,\beta)\, dv_s = \sum_{m=0}^{N} t^m \int_M \frac{1}{(4\pi t)^{n/2}} e^{-\rho^2/4t} \psi(q,s) u_m(s,q) \psi(s,\beta)\, dv_s \tag{42}$$

The integral of Eq. (42) over M can be broken up into an integral over $Q_q(\frac{\epsilon}{2}) = \{s \in M | \rho(q,s) < \epsilon/2\}$ anda second integral over the set $M - M_q(\frac{\epsilon}{2})$. On the latter set, the limit converges uniformly hence

$$\lim_{t\to\infty} \frac{e^{-\rho^2/4t}}{(4\pi t)^{n/2}} = 0$$

To estimate the remaining integral, choose a normal coordinate system at q and denote the integration coordinates as $(s_1,\ldots,s_n)$, then the integrand of Eq. (42) is given as

$$\frac{1}{(4\pi t)^{n/2}} e^{-|s|^2/4t} \varphi(q,s) u_m(s,q) \psi(s,\beta) \sqrt{\det\langle \frac{\partial}{\partial s_i}, \frac{\partial}{\partial s_j} \rangle}\, ds_1 \cdots ds_n$$

Therefore, in the limit using Definition 2.4,

$$\lim_{t\to 0} \int_{M(\epsilon/2)} \frac{1}{(4\pi t)^{n/2}} e^{-\rho^2/4t} \varphi(q,s) u_m(s,q) \psi(s,\beta)\, dv_s = u_m(q,q)\psi(q,\beta)$$

This result implies that

http://dx.doi.org/10.5772/64611

$$\lim_{t\to 0}\int_M G_0(t,q,s)\psi(s,\beta)\,dv_s = \sum_{m=0}^{N}\lim_{t\to 0} t^m u_m(q,q)\psi(q,\beta) = \psi(q,\beta)u_0(q,q) = \psi(q,\beta) \tag{43}$$

The convergence here is uniform.

There exists an asymptotic expansion for the heat kernel which is extremely useful and has several applications. It is one of the main intentions here to present this. An application of its use appears later.

Theorem 3.2. (Asymptotic expansion) Let M be a Riemannian manifold with dimension n and$E$ a vector bundle over M with inner product and metric Riemannian connection. Let $G(t,q,p)$ be the heat kernel or fundamental solution for heat operator (Eq. (15)) and (Eq. (39)) the MP parametrix. Then as $t\to 0$, $G(t,p,p)$ has the asymptotic expansion $G(t,p,p)\sim H\infty(t,p,p)$,

that is, for any $N>0$, it is the case that

$$G(t,p,p)-\frac{1}{(4\pi t)^{n/2}}\sum_{m=0}^{N} t^m u_m(p,p) = O\left(t^{N-\frac{n}{2}}\right) \tag{44}$$

and the symbol on the right-hand side of Eq. (44) signifies a quantity ξ with the property that

$$\lim_{t\to 0}\frac{\xi}{t^{N-\frac{n}{2}}} = 0$$

Proof: It suffices to prove the theorem for any large N. Let $G_0(t,q,p) = \varphi(q,p)H_N(t,q,p)$ as in Theorem 3.2. The conclusion of the theorem is equivalent to the statement

$$G(t,p,p)-G_0(t,p,p) = O\left(t^{N-\frac{n}{2}}\right)$$

From the previous theorem and existence and regularity of the fundamental solution, the result G of Levi iteration initialized by G_0 is exactly the fundamental solution. Equality (Eq. (41)) means that there exists a constant A such that for any $t\in(0,T)$,

$$|K_0(t,q,p)| = |\left(\frac{\partial}{\partial t}+\Delta\right)G_0(t,q,p)| \leq At^{N-\frac{n}{2}}$$

Let $v(M)$ be the volume of the manifold M. Using this result, the following upper bound is obtained

$$|K_1(t,q,p)| \leq \int_0^t d\tau\int_M |K_0(t-\tau,q,s)K_0(\tau,s,p)|\ dv_s$$

$$\leq \int_0^t [A^2(t-\tau)^{N-\frac{n}{2}}\tau^{N-\frac{n}{2}}v(M)]\,d\tau \leq \int_0^t A^2T^{N-\frac{n}{2}}\tau^{N-\frac{n}{2}}v(M)\,d\tau \leq AB\frac{t^{N-\frac{n}{2}+1}}{N-\frac{n}{2}+1}$$

We have set $B = A\cdot T^{N-\frac{n}{2}}v(M)$. Exactly the same procedure applies to $|K_2(t,q,p)|$. Based on the pattern established this way, induction implies that the following bound results

$$|K_m(t,q,p)| \le A \cdot B^m \frac{t^{N-\frac{n}{2}+m}}{(N-\frac{n}{2}+1)(N-\frac{n}{2}+2)\cdots(N-\frac{n}{2}+m)} \le A \cdot B^m \frac{t^m}{m!} t^{N-\frac{n}{2}}$$

The formula for Levi iteration yields upon summing this over m the following upper bound

$$|\tilde{K}(t,q,p)| \le \sum_{m=0}^{\infty} |K_m(t,q,p)| \le A \cdot e^{Bt} t^{N-\frac{n}{2}}$$

Using this bound, the required estimate is obtained,

$$|G(t,q,p)-G_0(t,q,p)| \le |\int_0^t d\tau \int_M dv_z G_0(t-\tau,q,z)\tilde{K}(\tau,z,p)|$$

$$\le \int_0^t d\tau \int_M \frac{e^{-\rho^2/4(t-\tau)}}{(4\pi(t-\tau))^{n/2}} A \cdot e^{B\tau} \cdot \tau^{N-\frac{n}{2}} dv_s$$

$$\le M_n A e^{Bt} \int_0^t \tau^{N-\frac{n}{2}} d\tau\, v(M) = \frac{1}{N-\frac{n}{2}+1} M_n A \cdot e^{Bt} v(M) t^{N-\frac{n}{2}+1}$$

This finishes the proof.

Now if all the Hodge theorem is used, formal expressions for the index can be obtained. Suppose $D : \Gamma(E) \to \Gamma(F)$ is an operator such that D^*D and DD^* are Schrödinger operators andD^* is the adjoint of D. Suppose the operators $D^*D : \Gamma(E) \to \Gamma(E)$ and $DD^* : \Gamma(F) \to \Gamma(E)$ are defined, so they are self-adjoint and have nonnegative real eigenvalues. Then the spaces $\Gamma_\mu(E)$ and $\Gamma_\mu(F)$ can be defined this way

$$\Gamma_\mu(E) = \{\varphi \in \Gamma(E)|D^*D\varphi = \mu\varphi\}, \qquad \Gamma_\mu(F) = \{\varphi \in \Gamma(F)|DD^*\varphi = \mu\varphi\} \tag{45}$$

For any $m > 0$, the dimensions of the spaces in (44) are finite and moreover,

$$\Gamma_0(E) = \ker\{D : \Gamma(E) \to \Gamma(F)\}, \qquad \Gamma_0(F) = \ker\{D^* : \Gamma(F) \to \Gamma(E)\}$$

Consequently, an expression for the index Ind (D) can be obtained from Eq. (45) as follows

$$\text{Ind } D = \dim \ker D - \dim \ker D^* = \dim \Gamma_0(E) - \dim \Gamma_0(F)$$

Definition 3.2. For the Schrödinger operator Δ, let $e^{-t\Delta} : \Gamma(E) \to \Gamma(E)$, for $t > 0$ be defined as

$$(e^{-t\Delta}\varphi)(q) = \int_M G(t,q,p)\varphi(p)dv_p \tag{46}$$

where $G(t,q,p)$ is the fundamental solution of heat operator (Eq. (15)).

Let $0 \le \lambda_1 \le \lambda_2 \le \cdots \to \infty$ be the eigenvalues of the operator Δ and $\{\psi_1,\psi_2,\ldots\}$ the corresponding eigenfunctions. Intuitively, the trace of $e^{-t\Delta}$ is defined as

$$tr\ e^{-t\Delta} = \sum_{k=1}^{\infty} \langle\langle e^{-t\Delta}\ \psi_k, \psi_k\rangle\rangle \tag{47}$$

This is clearly $\sum_k e^{-\lambda_k t}$ or $\sum_\mu e^{-t\mu}$ dim $\Gamma_\mu(E)$, so the definition of tr is well-defined if and only if

$$\sum_k e^{-\lambda_k t} < \infty \tag{48}$$

Theorem 3.3. For any $p,q \in M$, let $\{e_1(p),\dots,e_N(p)\}$ and $\{f_1(q),\dots,f_N(q)\}$ be orthonormal bases on E_p and E_q, respectively, then the following two results hold for $t > 0$,

$$\begin{aligned} &(a) \qquad \int_M \int \sum_{a,b=1}^{N} \langle G(t,q,p)e_a(p), f_b(q)\rangle^2\ dv_q dv_p < \infty, \\ &(b) \qquad \sum_{k=1}^{\infty} e^{2\lambda_k t} < \int_M \int \sum_{a,b=1}^{N} \langle G(t,q,p)e_a(p), f_b(q)\rangle^2\ dv_q dv_p < \infty \end{aligned} \tag{49}$$

Proof: When $t > 0$, $G(t,q,p)$ is continuous and hence satisfies (a). For and $w \in \Gamma(E)$, Theorem 2.5 yields the following expansion for $G(t,q,p) \in \overline{\Gamma(E)}$, hence the Parseval equality yields

$$\int_M |G(t,q,p)w|^2\ dv_q = \sum_{k=1}^{\infty} e^{-2\lambda_k t} \langle \psi_k(p), w\rangle^2$$

Replacing w by the basis element $e_a(p)$, this implies that

$$\begin{aligned} &\sum_{a=1}^{N} \int_M |G(t,q,p)e_a(p)|^2\ dv_q \\ &= \sum_{a=1}^{N} \sum_{k=1}^{\infty} e^{-2\lambda_k t} \langle \psi_k(p), e_a(p)\rangle^2 = \sum_{k=1}^{\infty} \sum_{a=1}^{N} e^{-2\lambda_k t}\ \langle \psi_k(p), e_a(p)\rangle^2 = \sum_{k=1}^{\infty} e^{-2\lambda_k t} \langle \psi_k(p), \psi_k(p)\rangle \end{aligned}$$

Then for any m, it follows that

$$\begin{aligned} &\sum_{k=1}^{m} e^{-2\lambda_k t} = \sum_{k=1}^{m} \int_M e^{-2\lambda_k t} \langle \psi_k(p), \psi_k(p)\rangle\ dv_p \leq \int_M \sum_{k=1}^{\infty} e^{-2\lambda_k t} \langle \psi_k(p), \psi_k(p)\rangle\ dv_p \\ &= \int_M dv_p \int_M \sum_{a=1}^{N} |G(t,q,p)e_a(p)|^2\ dv_q = \int_M \int_M \sum_{a,b=1}^{N} \langle G(t,q,p)e_a(p), f_b(q)\rangle^2\ dv_q dv_p < \infty \end{aligned}$$

Theorem 3.4. For any $t > 0$,

$$\text{tr}\ (e^{-t\Delta}) = \int_M tr\ G(t,p,p)\ dv_p \tag{50}$$

Proof: From Theorem 2.2, it follows that

$$tr\ G(t,p,p) = \sum_{a=1}^{N} \langle G(t,p,p)e_a(p),e_a(p)\rangle = \sum_{a=1}^{N} \left\langle \sum_{k=1}^{\infty} e^{-t\lambda_k} \langle \psi_k(p)e_a(p)\rangle \psi_k(p),e_a(p) \right\rangle$$

$$= \sum_{a=1}^{N} \sum_{k=1}^{\infty} e^{-t\lambda_k} \langle \psi_k(p),e_a(p)\rangle^2 = \sum_{k=1}^{\infty} e^{-t\lambda_k} \langle \psi_k(p),\psi_k(p)\rangle^2$$

Integrating this on both sides, it is found that

$$\int_M \ \mathrm{tr}\ G(t,p,p)\ dv_p = \int_M \sum_{k=1}^{\infty} e^{-t\lambda_k} \langle \psi_k(p),\psi_k(p)\rangle^2\ dv_p = \sum_{k=1}^{\infty} e^{-t\lambda_k} = \ \mathrm{tr}\ (e^{-t\Delta})$$

Note that Eq. (48) is a series with positive terms which converges uniformly as $t \to \infty$. Therefore,

$$\lim_{t\to\infty}\ \mathrm{tr}\ e^{-t\Delta} = \sum_{k=1}^{\infty} \lim_{t\to\infty}\ e^{-t\lambda_k} = \dim\ \Gamma_0(E) \tag{51}$$

In fact, as $t \to 0$, the equality

$$G(t,p,p) = \frac{1}{(4\pi t)^{n/2}} + O\left(\frac{1}{t^{n/2}}\right)$$

and the previous theorem imply that $\lim_{t\to 0}\ \mathrm{tr}\ e^{-t\Delta} = \infty$.

4. An application of the expansions: the Gauss Bonnet theorem

As far as $\mathrm{Ind}\ (D)$ is concerned, it is the case for all $t > 0$ that,

$$\mathrm{Ind}\ (D) = \ \mathrm{tr}\ e^{-tD^*D} - \ \mathrm{tr}\, e^{-tDD^*} = \int_M \ \mathrm{tr}\ G_+(t,p,p)\ dv_p - \int_M \ \mathrm{tr}\ G_-(t,p,p)\ dv_p$$

by Theorem 3.5, where $G_\pm(t,p,p)$ are the fundamental solutions of $\partial_t + D^*D$ and $\partial_t + DD^*$. As $t \to 0$, Theorem 3.2 assumes the form

$$G_\pm(t,p,p) \sim H_\infty^\pm(t,p,p) = \ \frac{1}{(4\pi t)^{n/2}} \sum_{m=0}^{\infty} t^m u_{\pm m}(p,p)$$

Lemma 4.1. Let $\{\lambda_i\}$ be the spectrum of the Laplacian on zero-forms, or functions, on M. Then,

$$\sum_k e^{-\lambda_k t} = \frac{1}{(4\pi t)^{n/2}} \sum_{k=0}^{\infty} \int_M u_k(x,x)\ dv_x \tag{52}$$

http://dx.doi.org/10.5772/64611

Proof:

$$\sum_k e^{-\lambda_k t} = \int_M \text{tr } G(t,x,x)\, dv_x = \frac{1}{(4\pi t)^{n/2}} \sum_k \left(\int_M u_k(x,x)\, dv_x \right) t^k$$

The spectrum of the Laplacian on functions characterizes a lot of interesting geometric information. Note that Eq. (52) can be written as

$$\sum_i e^{\lambda_i t} \sim \frac{1}{(4\pi t)^{n/2}} \sum_{k=0}^{\infty} a_k\, t^k, \qquad a_k = \int_M u_k(x,x)\, dv_x$$

and the trace does not appear in the case of functions. The superscript on the Laplacian Δ^p denotes the form degree acted upon andsimilarly on other objects throughout this section.

Two Riemannian manifolds are said to be isospectral if the eigenvalues of their Laplacians on functions counted with multiplicities coincide.

Corollary 4.1. Let M and N be compact isospectral Riemannian manifolds. Then M and N have the same dimension and the same volume.

Proof: Let $\{\lambda_i\}$ denote the spectrum of both M and N with $\dim M = m$ and $\dim N = n$. Then it follows that

$$\frac{1}{(4\pi t)^{m/2}} \sum_{k=0}^{\infty} \left(\int_M u_k^M(p,p)\, dv_p \right) t^k = \sum_{i=0}^{\infty} e^{-\lambda_i t} = \frac{1}{(4\pi t)^{n/2}} \sum_{k=0}^{\infty} \left(\int_N u_k^N(q,q)\, dv_q \right) t^k$$

This implies that $m = n$, which in turn implies that

$$\frac{1}{(4\pi t)^{m/2}} \left[\int_M u_0^M(p,p)\, dv_p - \int_N u^N(q,q)\, dv_q \right] = \frac{1}{(4\pi t)^{m/2}} \sum_{k=1}^{\infty} \left(\int_M u_k^M(p,p)\, dv_p - \int_N u^N(q,q)\, dv_q \right) t^k$$

Since the right-hand side of the equation depends on t, but the left-hand side does not, this result implies that

$$\int_M u_0^M(p,p)\, dv_p = \int_N u_0^N(q,q)\, dv_q \tag{53}$$

Iterating this argument leads to the set of equations

$$\int_M u_k^M(p,p)\, dv_p = \int_N u_k^N(q,q)\, dv_q \tag{54}$$

for all $k > 0$. In particular, since $u_0 = 1$, Eq. (53) leads to the conclusion vol $(M) =$ vol (N).

The proof illustrates that in fact there exist an infinite sequence of obstructions to claiming that two manifolds are isospectral, namely the set of integrals $\int_M u_k \, dv_p$. The first integral contains basic geometric information. It is then natural to investigate the other integrals in sequence as well. Recall that $R_p, \nabla R_p, \cdots$ denote the covariant derivatives of the curvature tensor at p. A polynomial P in the curvature and its covariant derivatives is called universal if its coefficients depend only on the dimension of M. The notation $P(R_p, \nabla R_p, \ldots, \nabla^k R_p)$ is used to denote a polynomial in the components of the curvature tensor and its covariant derivatives calculated in a normal Riemannian coordinate chart at p. The following theorem will not be proved, but it will be used shortly.

Theorem 4.2. On a manifold of dimension n,

$$u_1(p,p) = P_1^n(R_p), \qquad u_k(p,p) = P_k^n(R_p, \nabla R_p, \ldots, \nabla^{2k-2} R_p), \qquad k \geq 2 \tag{55}$$

for some universal polynomials P_k^n.

Thus, P_1^n is a linear function with no constant term and $u_1(p,p)$ is a linear function of the components of the curvature tensor at p, with no covariant derivative terms. The only linear combination of curvature components that produces a well-defined function $u_1(p,p)$ on a manifold is the scalar curvature $R(p) = R_{ij}^{ij}$ andso there exists a constant C such that $u_1(p,p) = C \cdot R(p)$.

Theorem 4.3.

$$u_1(p,p) = \frac{1}{6} R(p) \tag{56}$$

Proof: The proof amounts to noticing that P_1^n is a universal polynomial, so it suffices to compute C over one kind of manifold. A good choice is to integrate over S^n with the standard metric and work it out explicitly in normal coordinates. It is found that $u_1(p,p) = n(n-1)/6$ andit is known that $R(p) = n(n-1)$ for all $p \in S^n$ andthis implies Eq. (56).

The large t or long-time behavior of the heat operator for the Laplacian on differential forms is then controlled by the topology of the manifold through the means of the de Rham cohomology. The small t or short-time behavior is controlled by the geometry of the asymptotic expansion. The combination of topological information has a geometric interpretation. This is made explicit by means of the Chern-Gauss-Bonnet theorem. The two-dimensional version of this theorem will be developed here.

These results can be summarized by the elegant formula

$$\sum_{k=0}^{\infty} e^{-\lambda_k t} = \frac{1}{(4\pi t)^{n/2}} \left\{ v(M) + \frac{1}{6} \int_M R(x) \, dv_x \cdot t + O(t^2) \right\}$$

where $v(M)$ is the volume of M.

http://dx.doi.org/10.5772/64611

Suppose that λ is positive and here we let E_λ^p denote the possibly trivial eigenspace of Δ on p-forms. If $\omega \in E_\lambda^p$ then it follows that $\Delta^{p+1} d\omega = d\Delta^p \omega = \lambda\, d\omega$, hence $d\omega \in E_\lambda^{p+1}$. Thus, a well-defined sequential ordering of the spaces can be established. If $\omega \in E_\lambda^p$ has the property that $d\omega = 0$, then $\lambda\omega = \Delta^p \omega = (\delta d + d\delta)\omega = d\,\delta\omega$. Therefore, since $\lambda \neq 0$, it is found that $\omega = d\left(\frac{1}{\lambda}\delta\omega\right)$. Thus, the sequence $0 \to E_\lambda^0 \to^d \cdots \to^d E_\lambda^n \to 0$ is exact. Since the operator $d + \delta$ is an isomorphism on $\oplus_k E_\lambda^{2k}$, it follows that

$$\sum_s (-1)^s \dim E_\lambda^s = 0 \tag{57}$$

Theorem 4.4. Let $\{\lambda_i^s\}$ be the spectrum of the operator Δ, then

$$\sum_s (-1)^s \sum_i e^{-\lambda_i^s t} = \sum_s (-1)^s \dim \ker \Delta^s. \tag{58}$$

Proof: By (57),

$$\sum_s (-1)^s \sum_k e^{-\lambda_k^s t} = \sum_s (-1)^s \sum{}' e^{-\lambda_i t}$$

The sum on the right $\sum'$ is only over eigenvalues such that $\lambda_i^p = 0$ and so

$$\sum{}' e^{-\lambda_i^p t} = \dim \ker \Delta^p.$$

This has the consequence that

$$\sum_p (-1)^p \ \operatorname{tr} e^{-t\Delta} = \sum_p (-1)^p \sum_k e^{-\lambda_k^p t} \tag{59}$$

is independent of the parameter t. This means that its large or long t behavior is the same as its short or small t behavior. To put it another way, the long-time behavior of $\operatorname{tr} e^{-t\Delta}$ is given by the de Rham cohomology, while the short-time behavior is dictated by the geometry of the manifold. Using the definition of the Euler characteristic, it follows that

$$\begin{aligned} \chi(M) &= \textstyle\sum_p (-1)^p \dim H_{dH}^p(M) = \sum_p (-1)^p \dim \ker \Delta^p = \sum_p (-1)^p \ \operatorname{tr} e^{-t\Delta^p} \\ &= \textstyle\sum_p (-1)^p \displaystyle\int_M \operatorname{tr}\ G(t,x,x)\, dv_x \end{aligned} \tag{60}$$

From the asymptotic expansion theorem, the following expression for $\chi(M)$ results

$$\chi(M) = \frac{1}{(4\pi t)^{n/2}} \sum_{k=0}^{\infty} \left(\int_M \sum_{s=0}^{n} \operatorname{tr}\ u_k^s(x,x)\, dv_x \right) t^k \tag{61}$$

The u_k^s in Eq. (61) are the coefficients in the asymptotic expansion for $\operatorname{tr}(e^{-t\Delta^s})$. Since $\chi(M)$ is independent of t, only the constant or t-independent term on the right-hand side of Eq. (61) can be nonzero. This implies the following important theorem.

Theorem 4.5. If the dimension of M is even, then

$$\frac{1}{(4\pi)^{n/2}} \int_M \sum_{s=0}^{n} (-1)^s \text{ tr } u_k^s(x,x)\, dv_x = \begin{cases} 0, & k \neq \frac{n}{2}; \\ \chi(M), & k = \frac{n}{2}. \end{cases} \tag{62}$$

Theorem 4.6. (Gauss-Bonnet) Let M be a closed oriented manifold with Gaussian curvature K and area measure da_M, then

$$\chi(M) = \frac{1}{2\pi} \int_M K\, da_M \tag{63}$$

Proof: By the last theorem and the fact that tr $u_k^p(x,x) =$ tr $u_k^{p-1}(x,x)$, it follows that

$$\begin{aligned} \chi(M) &= \frac{1}{4\pi} \int_M \sum_{p=0}^{2} (^{-1)p} \text{ tr } u_1^p\, da_M = \frac{1}{4\pi} \int_M (\text{tr } u_1^0 - \text{ tr } u_1^1 + \text{ tr } u_1^2)\, da_M \\ &= \frac{1}{4\pi} \int_M (2 \text{ tr } u_1^0 - \text{ tr } u_1^1)\, da_M = \frac{1}{4\pi} \int_M (\frac{2}{3}K - \text{ tr } u_1^1)\, da_M \end{aligned} \tag{64}$$

since the scalar curvature is two times the Gaussian. Now it must be that tr $u_1^1(x,x) = CR(x) = 2CK(x)$, for some constant C. The standard sphere S^2 has Gaussian curvature one andso C can be calculated from Eq. (64),

$$2 = \frac{1}{2\pi} \int_{S^2} (\frac{1}{3} - C)\, da_M = \frac{1}{2\pi} (\frac{1}{3} - C) \cdot (4\pi)$$

Therefore, $C = -2/3$ and putting all of these results into Eq. (64), Eq. (62) results.

As an application of this theorem, note that the calculation of u_1 gives another topological obstruction to manifolds having the same spectrum.

Theorem 4.7. Let (M,g) and (N,h) be compact isospectral surfaces, then M and N are diffeomorphic.

Proof: As noted in Corollary 4.1,

$$\int_M u_1^M(x,x)\, dv_x = \int_N u_1^N(y,y)\, dv_y$$

On a surface, the scalar curvature is twice the Gaussian curvature, so by the Gauss-Bonnet theorem,

$$6\pi\chi(M) = \int_M u_1^M(x,x)\, dv_x = \int_N u_1^N(y,y)\, dv_y = 6\pi\chi(N) \tag{65}$$

However, oriented surfaces with the same Euler characteristic are diffeomorphic.

5. Summary and outlook

The heat equation approach has been seen to be quite deep, leading both to the Hodge theorem and also to a proof of the Gauss-Bonnet theorem. Moreover, it is clear from the asymptotic development that there is a generalization of this theorem to higher dimensions. The four-dimensional Chern-Gauss-Bonnet integrand is given by the invariant $\frac{1}{32\pi^2}\{K^2-4|\rho_r|^2+|R|^2\}$, where K is the scalar curvature, $|\rho_r|^2$ is the norm of the Ricci tensor, $|R|^2$ is the norm of the total curvature tensor andthe signature is Riemannian. This comes up in physics especially in the study of Einstein-Gauss-Bonnet gravity where this invariant is used to get the associated Euler-Lagrange equations.

Let R_{ijkl} be the components of the Riemann curvature tensor relative to an arbitrary local frame field $\{e_i\}$ for the tangent bundle TM and adopt the Einstein summation convention. Let $m=2s$ be even, then the Pfaffian $E_m(g)$ is defined to be

$$E_m(g)=\frac{1}{(8\pi)^s s!}\,R_{i_1 i_2 j_2 j_1}\cdots R_{i_{2s-1} i_{2s} j_{2s} j_{2s-1}}\,g(e^{i_1}\wedge\cdots\wedge e^{i_{2s}},e^{j_1}\wedge\cdots\wedge e^{j_{2s}}) \tag{66}$$

The Euler characteristic $\chi(M)$ of any compact manifold of odd dimension without boundary vanishes. Only the even dimensional case is of interest.

Theorem 5.1. Let (M,g) be a compact Riemannian manifold without boundary of even dimension m. Then

$$\chi(M)=\int_M E_m(g)\,dv_M \tag{67}$$

This was proved first by Chern, but of greater significance here, this can be deduced from the heat equation approach that has been introduced here. There is a proof by Patodi [18], but there is no room for it now. It should be hoped that more interesting results will come out in this area as well in the future.

Author details

Paul Bracken

Address all correspondence to: paul.bracken@utrgv.edu

Department of Mathematics, University of Texas, Edinburg, TX, USA

References

[1] Jost J. Riemannian Geometry and Geometric Analysis, Springer-Verlag, Berlin-Heidelberg; 2011.

[2] Yu Y. The Index Theorem and the Heat Equation Method, World-Scientific Publishing, Singapore; 2001.

[3] Berline N, Getzler E, Vergne M. Heat Kernel and Dirac Operators, Springer-Verlag, Berlin-Heidelberg, 1992.

[4] Rosenberg S. The Laplacian on a Riemannian Manifold, London Mathematical Society, 31, Cambridge University Press, New York, NY, USA; 1997.

[5] Gilkey P B. The Index Theorem and Heat Equation, Mathematics Lecture Series, No. 4, Publish or Perish Inc, Boston, MA; 1974.

[6] Goldberg S I. Curvature and Homology, Dover, New York, 1970.

[7] Cavicchidi A, Hegenbarth F. On the effective Yang-Mills Lagrangian and its equation of motion, J. Geom. Phys. 1998; **25**, 69–90.

[8] McKean, H Singer I. Curvature and eigenvalues of the Laplacian, J. Diff. Geom. 1967; **1**, 43–69.

[9] Atiyah M F, Patodi V K, Singer I. Spectral asymmetry and Riemannian geometry I, Math. Proc. Camb. Phil. Soc. 1975; 77, 43–69.

[10] Gilkey P B. Curvature and eigenvalues of Laplacian for elliptic complexes. Adv. Math. 1973; **11**, 311–325.

[11] Bracken P. Some eigenvalue bounds for the Laplacian on Riemannian manifolds. Int. J. Math Sciences. 2013; **8**, 221–226.

[12] Bracken P. The Hodge-de Rham decomposition theorem and an application to a partial differential equation. Acta Mathematica Hungarica, 2011; **133**, 332–341.

[13] Bracken P. A result concerning the Laplacian of the shape operator on a Riemannian manifold and an application. Tensor N S. 2013; **74**, 43–47.

[14] Minakshisundaram S, Pleijel A. Some properties of the eigenfunctions of the Laplace operator on Riemannian manifolds, Can J. Math. 1949; **1**, 242–256.

[15] Bracken P. A note on the fundamental solution of the heat operator on forms, Missouri J. Math Sciences. 2013; **25**, 186–194.

[16] Chern S S. A simple proof of the Gauss-Bonnet formula for closed Riemannian manifolds. Ann. Math. 1944; **45**, 747–752.

[17] Gilkey P B, Park J H. Analytic continuation, the Chern-Gauss-Bonnet theorem andthe Euler-Lagrange equations in Lovelock theory for indefinite signature metrics, J. Geom. Phys. 2015; **88**, 88–93.

[18] Patodi V K. Curvature and the eigenforms of the Laplace operator. J. Differen. Geom.. 1971; **5** 233–249.

Chapter 5

Symplectic Manifolds: Gromov-Witten Invariants on Symplectic and Almost Contact Metric Manifolds

Yong Seung Cho

Additional information is available at the end of the chapter

http://dx.doi.org/10.5772/65663

Abstract

In this chapter, we introduce Gromov-Witten invariant, quantum cohomology, Gromov-Witten potential, and Floer cohomology on symplectic manifolds, and in connection with these, we describe Gromov-Witten type invariant, quantum type cohomology, Gromov-Witten type potential and Floer type cohomology on almost contact metric manifolds. On the product of a symplectic manifold and an almost contact metric manifold, we induce some relations between Gromov-Witten type invariant and quantum cohomology and quantum type invariant. We show that the quantum type cohomology is isomorphic to the Floer type cohomology.

Keywords: symplectic manifold, Gromov-Witten invariant, quantum cohomology, Gromov-Witten potential, Floer cohomology, almost contact metric manifold, Gromov-Witten type invariant, quantum type cohomology, Gromov-Witten type potential, Floer type cohomology

1. Introduction

The symplectic structures of symplectic manifolds (M, ω, J) are, by Darboux's theorem 2.1, locally equivalent to the standard symplectic structure on Euclidean space.

In Section 2, we introduce basic definitions on symplectic manifolds [1–5, 10–13] and flux homomorphism. In Section 2.1, we recall J-holomorphic curve, moduli space of J-holomorphic curves, Gromov-Witten invariant and Gromov-Witten potential, quantum product and quantum cohomology, and in Section 2.2, symplectic action functional and its gradient flow line, Maslov type index of critical loop, Floer cochain complex and Floer cohomology, and theorem of Arnold conjecture.

In Section 3, we introduce almost contact metric manifolds $(M, g, \varphi, \eta, \xi, \phi)$ with a closed fundamental 2-form ϕ and their product [4, 7, 8]. In Section 3.1, we study φ-coholomorphic map, moduli space of φ-coholomorphic maps which represent a homology class of dimension two, Gromov-Witten type cohomology, quantum type product and quantum type cohomology, Gromov-Witten type potentials on the product of a symplectic manifold, and an almost contact metric manifold [5, 6, 13]. In Section 3.2, we investigate the symplectic type action functional on the universal covering space of the contractible loops, its gradient flow line, the moduli space of the connecting flow orbits between critical loops, Floer type cochain complex, and Floer type cohomology with coefficients in a Novikov ring [7, 9, 13].

In Section 4, as conclusions we show that the Floer type cohomology and the quantum type cohomology of an almost contact metric manifold with a closed fundamental 2-form are isomorphic [7, 13], and present some examples of almost contact metric manifolds with a closed fundamental 2-form.

2. Symplectic manifolds

By a symplectic manifold, we mean an even dimensional smooth manifold M^{2n} together with a global 2-form ω which is closed and nondegenerate, that is, the exterior derivative $d\omega = 0$ and the n-fold wedge product ω^n never vanishes.

Examples: (1) The $2n$-dimensional Euclidean space $\mathbb{R}^{2n}$ with coordinates $(x_1, \ldots, x_n, y_1, \ldots, y_n)$ admits symplectic form $\omega_0 = \sum_{i=1}^{n} dx_i \wedge dy_i$.

(2) Let M be a smooth manifold. Then its cotangent bundle T^*M has a natural symplectic form as follows. Let $\pi : T^*M \to M$ be the projection map and $x_1, \ldots, x_n$ are local coordinates of M. Then $q_i = x_i \circ \pi, i = 1, 2, \ldots, n$ together with fiber coordinates $p_1, \ldots, p_n$ give local coordinates of T^*M. The natural symplectic form on T^*M is given by

$$\omega = \sum_{i=1}^{n} dq_i \wedge dq_j. \tag{1}$$

(3) Every Kähler manifold is symplectic.

Darboux's Theorem 2.1 ([6]). *Every symplectic form ω on M is locally diffeomorphic to the standard form ω_0 on $\mathbb{R}^{2n}$.*

A symplectomorphism of (M, ω) is a diffeomorphism $\phi \in Diff(M)$ which preserves the symplectic form $\phi^*\omega = \omega$. Denote by $\mathrm{Sym}(M)$ the group of symplectomorphims of M. Since ω is nondegenerate, there is a bijection between the vector fields $X \in \Gamma(TM)$ and 1-forms $\omega(X, \cdot) \in \Omega^1(M)$. A vector field $X \in \Gamma(TM)$ is called symplectic if $\omega(X, \cdot)$ is closed.

Let M be closed, i.e., compact and without boundary. Let $\phi : \mathbb{R} \to \mathrm{Diff}(M)$, $t \mapsto \phi_t$ be a smooth family of diffeomorphisms generated by a family of vector fields $X_t \in \Gamma(TM)$ via,

$$\frac{d}{dt}\phi_t = X_t \circ \phi_t, \phi_0 = \text{id}. \tag{2}$$

Then $\phi_t \in \text{Symp}(M)$ if and only if $X_t \in \Gamma(TM, \omega)$ the space of symplectic vector fields on M. Moreover, if $X,Y \in \Gamma(TM, \omega)$, then $[X,Y] \in \Gamma(TM, \omega)$ and $\omega([X, Y], \cdot) = dH$, where $H = \omega(X, Y) : M \to R$. Let $H : M \to R$ be a smooth function. Then the vector field X_H on M determined by $\omega(X_H, \cdot) = dH$ is called the Hamiltonian vector field associated with H. If M is closed, then X_H generates a smooth 1-parameter group of diffeomorphisms $\phi_H^t \in \text{Diff}(M)$ such that

$$\frac{d}{dt}\phi_H^t = X_H \circ \phi_H^t, \phi_H^0 = \text{id}. \tag{3}$$

This $\{\phi_H^t\}$ is called the Hamiltonian flow associated with H. The flux homomorphism Flux is defined by

$$\text{Flux}\{\phi_H^t\} = \int_0^1 \omega(X_t, \cdot)dt. \tag{4}$$

Theorem 2.2 ([6]). *$\phi \in Sym(M)$ is a Hamiltonian symplectomorphism if and only if there is a homotopy $[0, 1] \to \text{Sym}(M)$, $t \mapsto \phi_t$ such that $\phi_0 = \text{id}$, $\phi_1 = \phi$, and $\text{Flux}(\{\phi_t\}) = 0$.*

2.1. Quantum cohomology

Let (M, ω) be a compact symplectic manifold. An almost complex structure is an automorphism of TM such that $J^2 = -I$. The form ω is said to tame J if $\omega(v, Jv) > 0$ for every $v \neq 0$. The set $\mathfrak{I}_\tau(M, \omega)$ of almost complex structures tamed by ω is nonempty and contractible. Thus the Chern classes of TM are independent of the choice $J \in \mathfrak{I}_\tau(M, \omega)$. A smooth map $\phi : (M_1, J_1) \to (M_2, J_2)$ from M_1 to M_2 is (J_1, J_2)-holomorphic if and only if

$$d\phi_x \circ J_1 = J_2 \circ d\phi_x \tag{5}$$

Hereafter, we denote by $H_2(M)$ the image of Hurewicz homomorphism $\pi_2 M \to H_2(M, \mathbb{Z})$. A (i, J)-holomorphic map $u : (\Sigma, z_1, \ldots, z_k) \to M$ from a reduced Riemann surface (Σ, j) of genus g with k marked points to (M, J) is said to be stable if every component of Σ of genus 0 (resp. 1), which is contracted by u, has at least 3 (resp. 1) marked or singular points on its component, and the k marked points are distinct and nonsingular on Σ. For a two-dimensional homology class $A \in H_2(M)$ let $\mathcal{M}_{g,k}(M, A; J)$ be the moduli space of (j, J)-holomorphic stable maps which represent A.

Let $B := C^\infty(\Sigma, M; A)$ be the space of smooth maps

$$u : \Sigma \to M \tag{6}$$

which represent $A \in H_2(M)$.

Let us consider infinite dimensional vector bundle $E \to B$ whose fiber at u is the space $E_u = \Omega^{0,1}(\Sigma, u^*TM)$ of smooth J-antilinear 1-forms on Σ with values in u^*TM. The map $\bar{\partial}_J : B \to E$ given by

$$\bar{\partial}_J(u) = \frac{1}{2}(du + J \circ du \circ j) \tag{7}$$

is a section of the bundle. The zero set of the section $\bar{\partial}_J$ is the moduli space $\mathcal{M}_{g,k}(M, A; J)$.

For an element $u \in \mathcal{M}_{g,k}(M, A; J)$ we denote by

$$D_u : \Omega^0(\Sigma, u^*TM) = T_u B \to \Omega^{0,1}(\Sigma, u^*TM) \tag{8}$$

the composition of the derivative

$$d(\bar{\partial}_J)_u : T_u B \to T_{(u,0)} E \tag{9}$$

with the projection to fiber $T_{(u,0)}E \to \Omega^{0,1}(\Sigma, u^*TM)$. Then the virtual dimension of $\mathcal{M}_{g,k}(M, A; J)$ is

$$\begin{aligned} \dim \mathcal{M}_{g,k}(M, A; J) &= \operatorname{index} D_u : \Omega^0(\Sigma, u^*TM) \to \Omega^{0,1}(\Sigma, u^*TM) \\ &= 2c_1(TM)A + n(2-2g) + (6g-6) + 2k. \end{aligned} \tag{10}$$

Theorem 2.1.1. *For a generic almost complex structure $J \in \Im_\tau(M, \omega)$ the moduli space $\mathcal{M}_{g,k}(M, A; J)$ is a compact stratified manifold of virtual dimension,*

$$\dim \mathcal{M}_{g,k}(M, A; J) = 2c_1(TM)A + n(2-2g) + (6g-g) + 2k. \tag{11}$$

For some technical reasons, we assume that $c_1(A) \geq 0$ if $\omega(A) > 0$ and A is represented by some J-holomorphic curves. In this case, we call the symplectic manifold M semipositive. We define the evaluation map by

$$\text{ev} : \mathcal{M}_{g,k}(M, A; J) \to M^k, \ \text{ev}([u; z_1, \ldots, z_k]) = (u(z_1), \ldots, u(z_k)). \tag{12}$$

Then the image Im(ev) is well defined, up to cobordism on J, as a $\dim \mathcal{M}_{g,k}(M, A; J) :\equiv m$-dimensional homology class in M^k.

Definition. The Gromov-Witten invariant $\Phi_{g,k}^{M,A}$ is defined by

$$\Phi_{g,k}^{M,A} : H^m(M^k) \to Q, \Phi_{g,k}^{M,A}(a) = \int_{\mathcal{M}_{g,k}(M,A;J)} \text{ev}^* \bullet \alpha \tag{13}$$

where $\alpha = PD(a) \in H_{2nk-m}(M^k)$ and $\bullet$ is the intersection number of ev and α in M^k.

The minimal Chern number N of (M, ω) is the integer $N := \min \ \}7B\lambda | c_1(A) = \lambda \geq 0, A \in H_2(M)\}$. We define the quantum product $a * b$ of $a \in H^k(M)$ and $b \in H \ \ (M)$ as the formal sum

http://dx.doi.org/10.5772/65663

$$a * b = \sum_{A \in H_2(M)} (a * b)_A q^{c_1(A)/N} \tag{14}$$

where q is an auxiliary variable of degree $2N$ and $(a * b)_A \in H^{k+l-2c_1(A)}(M)$ is defined by

$$\int_C (a * b)_A = \Phi_{0,3}^{M,A}(a \otimes b \otimes r) \tag{15}$$

for $C \in H_{k+l-2c_1(A)}(M)$, $r = PD(C)$. Hereafter, we use the Gromov-Witten invariants of $g = 0$ and $k = 3$. Then the quantum product $a * b$ is an element of

$$QH^* := H^*(M) \otimes \mathbb{Q}[q] \tag{16}$$

where $\mathbb{Q}[q]$ is the ring of Laureut polynomials of the auxiliary variable q.

Extending $*$ by linearity, we get a product called quantum product

$$* : QH^*(M) \otimes QH^*(M) \to QH^*(M). \tag{17}$$

It turns out that $*$ is distributive over addition, skew-commutative, and associative.

Theorem 2.1.2. *Let (M, ω) be a compact semipositive symplectic manifold. Then the quantum cohomology $(QH^*(M), +, *)$ is a ring.*

Remark. For $A = 0 \in H_2(M)$, the all J-holomorphic maps in the class A are constant. Thus $(a * b)_0 = a \cup b$. The constant term of $a * b$ is the usual cup product $a \cup b$.

We defined the Novikov ring Λ_ω by the set of functions $\lambda : H_2(M) \to Q$ that satisfy the finiteness condition

$$\# \{A \in H_2(M) | \lambda(A) \neq 0, \omega(A) < c\} < \infty \tag{18}$$

for every $c \in R$. The grading is given by $\deg(A) = 2c_1(A)$.

Examples ([5]). (1) Let $p \in H^2(\mathbb{C}P^n)$ and $A \in H_2(\mathbb{C}P^n)$ be the standard generators. There is a unique complex line through two distinct points in $\mathbb{C}P^n$ and so $p * p^n = q$. The quantum cohomology of $\mathbb{C}P^n$ is

$$QH^*(\mathbb{C}P^n; \mathbb{Q}[q]) = \frac{\mathbb{Q}[p, q]}{< p^{n+1} = q >}. \tag{19}$$

(2) Let $G(k, n)$ be the Grassmannian of complex k-planes in $\mathbb{C}^n$. There are two natural complex vector bundles $\mathbb{C}^k \to E \to G(k, n)$ and $\mathbb{C}^{n-k} \to F \to G(k, n)$. Let $x_i = c_i(E^*)$ and $y_i = c_i(F^*)$ be Chern classes of the dual bundles E^* and F^*, respectively. Since $E \oplus F$ is trivial, $\sum_{i=0}^{j} x_i y_{j-i} = 0, j = 1, \ldots, n$. By computation $x_k * y_{n-k} = (-1)^{n-k} q$. The quantum cohomology of $G(k, n)$ is

$$QH^*\Big(G(k,n);\mathbb{Q}[q]\Big) = \frac{\mathbb{Q}[x_1,\ldots,x_k,q]}{< y_{n-k+1},\ldots,y_{n-1},y_n + (-1)^{n-k}q >}. \tag{20}$$

Let $\{e_0,\ldots,e_n\}$ be an integral basis of $H^*(M)$ such that $e_0 = 1 \in H^0(M)$ and each e_i has pure degree. We introduce $n+1$ formal variables $t_0,\ldots,t_n$ and the linear polynomial a_t in $t_0,\ldots,t_n$ with coefficients in $H^*(M)$ by $a_t = t_0e_0 + \cdots + t_ne_n$. The Gromov-Witten potential of (M,ω) is a formal power series in variables $t_0,\ldots,t_n$ with coefficients in the Novikov ring Λ_ω

$$\begin{aligned}\Phi^M(t) &= \sum_{k\geq 3}\sum_{A}\frac{1}{k!}\Phi^{M,A}_{0,k}(a_t,\ldots,a_t)q^{\frac{c_1(A)}{N}} \\ &= \sum_{k_0+\cdots+k_n\geq 3}\sum_{A}\frac{\varepsilon(k_0,\ldots,k_n)}{k_0!\ldots k_n!}\Phi^{M,A}_{0,k}(e_0^{k_0}\otimes\cdots\otimes e_n^{k_n})\cdot(t^0)k_0\ldots(t^n)^{k_n}q^{c_1(A)/N}.\end{aligned} \tag{21}$$

Examples ([4]). (1) $\Phi^{\mathbb{C}P^1}(t) = \frac{1}{2}t_0{}^2t_1 + \left(e^{t_1} - 1 - t_1 - \frac{t_1^2}{2}\right)$.

(2) $\Phi^{\mathbb{C}P^n}(t) = \frac{1}{6}\sum_{i+j+k=n} t_it_jt_k + \sum_{d>0}\sum_{k_2\ldots k_n} N_d(k_2\ldots k_n)\cdot\frac{t_2^{k_2}\ldots t_n^{k_n}}{k_2!\ldots k_n!}e^{dt_1}q^d$,

where $N_d(k_2\ldots k_n) = \Phi^{\mathbb{C}P^n, dA}_{0,k}(p^2\ldots p^2,\ldots,p^n\ldots p^n)$.

We define a nonsingular matrix (g_{ij}) by $g_{ij} = \int_M e_i \cup e_j$ and denote by (g^{ij}) its inverse matrix.

Theorem 2.1.3 ([4, 5]). *The Gromov-Witten potential Φ^M of (M,ω) satisfies the WDVV-equations:*

$$\sum_{\upsilon,\mu}\partial_{t_i}\partial_{t_j}\partial_{t_\upsilon}\Phi^M(t)g^{\upsilon\mu}\partial_{t_\mu}\partial_{t_k}\partial_{t}\ \Phi^M(t) = \varepsilon_{ijk}\cdot\sum_{\upsilon,\mu}\partial_{t_j}\partial_{t_k}\partial_{t_\upsilon}\Phi^M(t)g^{\upsilon\mu}\partial_{t_\mu}\partial_{t_i}\partial_{t}\ \Phi^M(t), \tag{22}$$

where $\varepsilon_{ijk} = (-1)^{\deg(e_1)(\deg(e_j)+\deg(e_k))}$.

2.2. Floer cohomology

Let a compact symplectic manifold (M,ω) be semipositive. Let $H_{t+1}: M \to R$ be a smooth 1-periodic family of Hamiltonian functions. The Hamiltonian vector field X_t is defined by $\omega(X_t,\cdot) = dH_t$. The solutions of the Hamiltonian differential equation $\dot{x}(t) = X_t(x(t))$ generate a family of Hamiltonian symplectomorphisms $\phi_t : M \to M$ satisfying $\frac{d}{dt}\phi_t = X_t \circ \phi_t$ and $\phi_0 = \mathrm{id}$. For every contractible loop $x : \mathbb{R}/\mathbb{Z} \to M$, there is a smooth map $u : D := $ }7Bz∈C||z|≤1} $\to M$ such that $u(e^{2\pi it}) = x(t)$. Two such maps u_1 and u_2 are called equivalent if their boundary sum$(u_1)\#(-u_2)$ is homologus to zero in $H_2(M)$. Denote by $(x,[u_1]) = (x,[u_2])$ for equivalent pairs, LM the space of contractible loops and $\tilde{LM}$ the space of equivalence classes. Then $\widetilde{LM} \to LM$ is a covering space whose covering transformation group is $H_2(M)$ via, $A(x,[u]) = (x,[A\#u])$ for each $A \in H_2(M)$ and $(x,[u]) \in \widetilde{LM}$.

Definition. The symplectic action functional a_H is defined by

$$a_H : \widetilde{LM} \to R, a_H(x,[u]) = -\int_D u^*\omega - \int_0^1 H_t\Big(x(t)\Big)dt. \tag{23}$$

For each element $\tilde{x} := (x,[u]) \in \widetilde{LM}$ and $\xi \in T_{\tilde{x}}\widetilde{LM}$, we have

$$da_H(x,[u])(\xi) = \int_0^1 \omega\Big(\dot{x}(t) - X_t\Big(x(t)\Big), \xi\Big)dt. \tag{24}$$

Thus the critical points of a_H are one-to-one correspondence with the periodic solutions of $\dot{x}(t) - X_t\Big(x(t)\Big) = 0$. Denote by $\tilde{PH} \subset \tilde{LM}$ the critical points of a_H and by $PH \subset LM$ the set of periodic solutions.

The gradient flow lines of a_H are the solutions $u : \mathbb{R}^2 \to M$ of the partial differential equation

$$\partial_u + J(u)\Big(\partial_t u - X_t(u)\Big) = 0$$

with conditions $u(s,t+1) = u(s,t)$,

$$\lim_{s\to\pm\infty} u(s,t) = x^{\pm}(t) \tag{25}$$

for some $x^{-} \in PH$.

Let $\mathcal{M}(\tilde{x}^-, \tilde{x}^+)$ be the space of such solutions u with $\tilde{x}^+ = \tilde{x}^- \# u$. This space is invariant under the shift $u(s,t) \mapsto u(s+s_0,t)$ for each $s_0 \in R$. For a generic Hamiltonian function, the space $\mathcal{M}(\tilde{x}^-, \tilde{x}^+)$ is a manifold of dimension

$$\dim \mathcal{M}(\tilde{x}^-, \tilde{x}^+) = \mu(\tilde{x}^-) - \mu(\tilde{x}^+). \tag{26}$$

Here $\mu : \tilde{PH} \to Z$ is a version of Maslov index defined by the path of symplectic matrices generated by the linearized Hamiltonian flow along $x(t)$.

Let $\mu(\tilde{x}) - \mu(\tilde{y}) = 1$. Then $\mathcal{M}(\tilde{x},\tilde{y})$ is a one-dimensional manifold and the quotient by shift $\mathcal{M}(\tilde{x},\tilde{y})/\mathbb{R}$ is finite. In this case, we denote by $n(\tilde{x},\tilde{y}) = \#\Big(\frac{\mathcal{M}(\tilde{x},\tilde{y})}{\mathbb{R}}\Big)$ the number of connecting orbits from $\tilde{x}$ to $\tilde{y}$ counted with appropriate signs.

We define the Floer cochain group $FC^*(M,H)$ as the set of all functions $\xi : \tilde{PH} \to Q$ that satisfy the finiteness condition,

$$\#\{\tilde{x} \in \widetilde{PH} \mid \xi(\tilde{x}) \neq 0, a_H(\tilde{x}) \leq c\} < \infty \tag{27}$$

for every $c \in R$. The complex $FC^*(M,H)$ is a Λ_ω-module with action

$$(\lambda * \xi)(\tilde{x}) := \sum_A \lambda(A)\xi(A\#\tilde{x}). \tag{28}$$

The degree k part $FC^*(M,H)$ consists of all $\xi \in FC^*(M,H)$ that are nonzero only on elements $\tilde{x} \in \widetilde{PH}$ with $\mu(\tilde{x}) = k$. Thus the dimension of $FC^*(M,H)$ as a Λ_ω-module is the number $\#(PH)$.

We define a coboundary operator $\delta : FC^*(M,H) \to FC^{k+1}(M,H)$ by

$$\delta(\xi)(\tilde{x}) = \sum_{\mu(\tilde{y})=k} \boldsymbol{n}(\tilde{x},\tilde{y})\xi(\tilde{y}). \tag{29}$$

The coefficients of $\delta(\delta(\xi)(\tilde{x}))$ are given by counting the numbers of pairs of connecting orbits from $\tilde{x}$ to $\tilde{y}$ where $\mu(\tilde{x}) - \mu(\tilde{y}) = 2 = \dim\mathcal{M}(\tilde{x},\tilde{y})$. The quotient $\mathcal{M}(\tilde{x},\tilde{y})/\mathbb{R}$ is a one-dimensional oriented manifold that consists of pairs counted by $\delta(\delta(\xi)(\tilde{x}))$. Thus the numbers cancel out in pairs, so that $\delta\left(\delta(\xi)\right) = 0$.

Definition. The cochain complex $(FC^*(M,H),\delta)$ induces its cohomology groups

$$FH^k(M,H) := \frac{\mathrm{Ker}\delta : FC^k(M,H) \to FC^{k+1}(M,H)}{\mathrm{Im}\delta : FC^{k-1}(M,H) \to FC^k(M,H)} \tag{30}$$

which are called the Floer cohomology groups of (M,ω,H,J).

Remark. By the usual cobordism argument, the Floer cohomology groups $FH^*(M,H)$ are independent to the generic choices of H and J. Let $f : M \to R$ be a Morse function such that the negative gradient flow of f with respect to the metric $g(\cdot,\cdot) = \omega(\cdot,J\cdot)$ is Morse-Smale. Let $H = -\varepsilon f : M \to R$ be the time-independent Hamiltonian. If ε is small, then the 1-periodic solutions of $\dot{x}(t) - X_H\left(x(t)\right) = 0$ are one-to-one correspondence with the critical points of f. Thus we have $PH = \mathrm{Crit}(f)$ and the Maslov type index can be normalized as

$$\mu(x,[u]) = \mathrm{ind}_f(x) - n \tag{31}$$

where $u : D \to M$ is the constant map $u(D) = x$.

We define a cochain complex $MC^*(M;\Lambda_\omega)$ as the graded Λ_ω-module of all functions

$$\xi : \mathrm{Crit}(f)H_2(M) \to Q \tag{32}$$

that satisfy the finiteness condition

$$\#\{(x,A)|\xi(x,A)\neq 0, \omega(A)\geq c\} < \infty \tag{33}$$

for every $c\in R$. The Λ_ω-module structure is given by $(\lambda * \xi)(x,A) = \sum \lambda(B)\xi(x,A+B)$ and the grading $\deg(x,A) = \mathrm{ind}_f(x) - 2c_1(A)$. The gradient flow lines $u : \mathbb{R} \to M$ of f are the solutions of $\dot{u}(s) = -\nabla f(u(s))$. These solutions determine coboundary operator

$$\delta : MC^k(M;\Lambda_\omega) \to MC^{k+1}(M;\Lambda_\omega) \tag{34}$$

$$\delta(\xi)(x,A) = \sum_y n_f(x,y)\xi(y,A) \tag{35}$$

where $n_f(x,y)$ is the number of connecting orbits u from x to y satisfying $\lim_{s\to-\infty} u(s) = x$, $\lim_{s\to+\infty} u(s) = y$, counted with appropriate signs and $\mathrm{ind}_f(x) - \mathrm{ind}_f(y) = 1$.

http://dx.doi.org/10.5772/65663

Definition–Theorem 2.2.1. (1) *The cochain complex* $(MC^*(M;\Lambda_\omega),\delta)$ *defines a cohomology group*

$$MH^*(M;\Lambda_\omega) := \frac{\mathrm{Ker}\delta : MC^*(M;\Lambda_\omega) \to MC^{*+1}(M;\Lambda_\omega)}{\mathrm{Im}\delta : MC^{*-1}(M;\Lambda_\omega) \to MC^*(M;\Lambda_\omega)} \tag{36}$$

which is called the Morse-Witten cohomology of M.

(2) $MH^*(M;\Lambda_\omega)$ *is naturally isomorphic to the quantum cohomology* $QH^*(M;\Lambda_\omega)$.

Theorem 2.2.2 ([5]). *Let a compact symplectic manifold* (M,ω) *be semipositive. There is an isomorphism*

$$\Phi : FH^*(M,H) \to QH^*(M;\Lambda_\omega) \tag{37}$$

which is linear over the Novikov ring Λ_ω.

Let $H : M \to R$ be a generic Hamiltonian function and $\phi : M \to M$ the Hamiltonian symplectomorphism of H. By Theorems 2.2.1 and 2.2.2

$$FH^*(M,H) \simeq QH^*(M;\Lambda_\omega) \simeq H^*(M) \otimes \Lambda_\omega \tag{38}$$

The rank of $FC^*(M,H)$ as a Λ_ω-module must be at least equal to the dimension of $H^*(M)$. The rank is the number $\#(PH)$ which is the number of the fixed points of ϕ.

Theorem 2.2.3 (Arnold conjecture). *Let a compact symplectic manifold* (M,ω) *be semipositive. If a Hamiltonian symplectomorphism* $\phi : M \to M$ *has only nondegenerate fixed points, then*

$$\#(\mathrm{Fix}(\phi)) \geq \sum_{j=0}^{2n} b_j(M) \tag{39}$$

where $b_j(M)$ *is the jth Betti number of M.*

3. Almost contact metric manifolds

Let be a real $(2n+1)$-dimensional smooth manifold. An almost cocomplex structure on M is defined by a smooth $(1,1)$ type tensor φ, a smooth vector field ξ, and a smooth 1-form η on M such that for each point $x \in M$,

$$\varphi_x^2 = -I + \eta_x \otimes \xi_x, \eta_x(\xi_x) = 1, \tag{40}$$

where $I : T_xM \to T_xM$ is the identity map of the tangent space T_xM.

A Riemannian manifold M with a metric tensor g and with an almost co-complex structure (φ, ξ, η) such that

$$g(X,Y) = g(\varphi X, \varphi Y) + \eta(X)\eta(Y), X, Y \in \Gamma(TM), \tag{41}$$

is called an almost contact metric manifold.

The fundamental 2-form ϕ of an almost contact metric manifold $(M, g, \varphi, \xi, \eta)$ is defined by

$$\phi(X, Y) = g(X, \varphi Y) \tag{42}$$

for all $X, Y \in \Gamma(TM)$. The $(2n+1)$-form $\phi^n \wedge \eta$ does not vanish on M, and so M is orientable. The Nijehuis tensor [8, 11] of the (1,1) type tensor φ is the (1,2) type tensor field N_φ defined by

$$N_\varphi(X, Y) = [\varphi X, \varphi Y] - [X, Y] - \varphi[\varphi X, Y] - \varphi[X, \varphi Y] \tag{43}$$

for all $X, Y \in \Gamma(TM)$, where $[X, Y]$ is the Lie bracket of X and Y. An almost cocomplex structure (φ, ξ, η) on M is said to be integrable if the tensor field $N_\varphi = 0$, and is normal if $N_\varphi + 2d\eta \otimes \xi = 0$.

Definition. An almost contact metric manifold $(M, g, \varphi, \eta, \xi, \phi)$ is said to be

1. almost cosymplectic (or almost co-Kähler) if $d\phi = 0$ and $d\eta = 0$,
2. contact (or almost Sasakian) if $\phi = d\eta$,
3. an almost C-manifold if $d\phi = 0$, $d\eta \neq 0$, and $d\eta \neq \phi$,
4. cosymplectic (co-Kähler) if M is an integrable almost cosymplectic manifold,
5. Sasakian if M is a normal almost Sasakian manifold,
6. a C-manifold if M is a normal almost C-manifold.

An example of compact Sasakian manifolds is an odd-dimensional unit sphere S^{2n+1}, and the one of the co-Kähler (almost cosymplectic) manifolds is a product MS^1 where M is a compact Kähler (symplectic) manifold, respectively.

Let $(M_1^{2n_1+1}, g_1, \varphi_1, \eta_1, \xi_1)$ and $(M_2^{2n_2+1}, g_2, \varphi_2, \eta_2, \xi_2)$ be almost contact metric manifolds. For the product $M := M_1 M_2$, Riemannian metric on M is defined by

$$g\Big((X_1, Y_1), (X_2, Y_2)\Big) = g_1(X_1, X_2) + g_2(Y_1, Y_2). \tag{44}$$

An almost complex structure on M is defined by

$$J(X, Y) = \Big(\varphi_1(X) + \eta_2(Y)\xi_1, \varphi_2(Y) - \eta_1(X)\xi_2\Big). \tag{45}$$

Then $J^2 = -I$ and the fundamental 2-form ϕ on M is $\phi = \phi_1 + \phi_2 + \eta_1 \wedge \eta_2$. If ϕ_1, ϕ_2 and η_1 and η_2 are closed, then ϕ is closed. Thus we have

Theorem 3.1. *Let $(M_1^{2n_1+1}, g_1, \varphi_1, \eta_1, \xi_1)$ be almost contact metric manifolds, $j = 1, 2$, and (M, g, ϕ, J) be the product constructed as above.*

1. *If ϕ_i and η_i, $i = 1,2$, are closed, then ϕ is closed.*
2. *J is an almost complex structure on M.*

http://dx.doi.org/10.5772/65663

3. *If M_i, $i = 1, 2$, are cosymplectic, then M is Kähler.*

Let $(M_1^{2n_1}, g_1, J_1)$ be a symplectic manifold, and $(M_2^{2n_2+1}, g_2, \varphi_2, \eta_2, \xi_2)$ be an almost contact metric manifold. Then $\xi_1 = \eta_1 = 0$, and $\omega_1 = \phi_1$ on M_1.

Theorem 3.2. *Let $(M, g, \varphi, \eta, \xi)$ be the product constructed as above.*

1. *If M_2 is contact, then M is an almost C-manifold.*

2. *If M_2 is a C-manifold, then M is an almost C-manifold.*

3. *If M_2 is almost cosymplectic, then M is almost cosymplectic.*

3.1. Quantum type cohomology

In [10, 11] we have studied the quantum type cohomology on contact manifolds. In this section, we want to introduce the quantum type cohomologies on almost cosymplectic, contact, and C-manifolds.

Let $(M^{2n+1}, g, \varphi, \eta, \xi)$ be an almost contact metric manifold. Then the distribution $\mathfrak{H} = \{X \in TM | \eta(X) = 0\}$ is an n-dimensional complex vector bundle on M.

Now fix the vector bundle $\mathfrak{H} \to M$. As the symplectic manifolds, a (1,1) type tensor field $\varphi : \mathfrak{H} \to H$ with $\varphi^2 = -I$ is said to be tamed by ϕ if $\phi(X, \varphi X) > 0$ for $X \in \mathfrak{H} \backslash \{0\}$ is said to be compatible if $\phi(\varphi X, \varphi Y) = \phi(X, Y)$.

Assume that the almost contact metric manifold M has a closed fundamental 2-form ϕ, i.e., $d\phi = 0$. An almost contact metric manifold M with the ϕ is called semipositive if for every $A \in \pi_2(M)$, $\phi(A) > 0$, $c_1(\mathfrak{H})(A) \geq 3-n$, then $c_1(\mathfrak{H})(A) > 0$ [13]. A smooth map $u : (\Sigma, j) \to (M, \varphi)$ from a Riemann surface (Σ, j) into (M, φ) is said to be φ-coholomorphic if $du \circ j = \varphi \circ du$.

Let $A \in H_2(M; \mathbb{Z})$ be a two-dimensional integral homology class in M. Let $\mathcal{M}_{0,3}(M; A, \varphi)$ be the moduli space of stable rational φ-coholomorphic maps with three marked points, which represent class A.

Lemma 3.1.1. *For a generic almost complex structure φ on the distribution, $\mathbb{C}^n \to H \to M$, the moduli space $\mathcal{M}_{0,3}(M; A, \varphi)$ is a compact stratified manifold with virtual dimension $2c_1(\mathfrak{H})[A] + 2n$.*

Consider the evaluation map given by

$$\mathrm{ev} : \mathcal{M}_{0,3}(M; A, \varphi) \to M^3, \tag{46}$$

$$\mathrm{ev}(\Sigma; z_1, z_2, z_3, u) = \Big(u(z_1), u(z_2), u(z_3)\Big). \tag{47}$$

We have a Gromov-Witten type invariant given by

$$\Phi_{0,3}^{M,A,\varphi} : H^*(M^3) \to Q \tag{48}$$

$$\Phi_{0,3}^{M,A,\varphi}(\alpha) = \int_{\mathcal{M}_{0,3}(M; A, \varphi)} \mathrm{ev}^*(\alpha) = \mathrm{ev}_*[\mathcal{M}_{0,3}(M; A, \varphi)] \cdot PD(\alpha) \tag{49}$$

which is the number of these intersection points counted with signs according to their orientations.

We define a quantum type product $*$ on $H^*(M)$, for $\alpha\in H^k(M)$ and $\beta\in H\ \ (M)$,

$$\alpha * \beta = \sum_{A\in H_2(M)} (\alpha * \beta)_A q^{c_1(\mathfrak{H})[A]/N}, \tag{50}$$

where N is called the minimal Chern number defined by

$$< c_1(\mathfrak{H}), H_2(M) >= N\mathbb{Z} \tag{51}$$

The $(\alpha * \beta)_A \in H^{k+l-2c_1(\mathfrak{H})[A]}(M)$ is defined for each $C\in H_{k+l-2c_1(\mathfrak{H})[A]}(M)$,

$$\int_C (\alpha * \beta)_A = \Phi_{0,3}^{M,A,\varphi}(\alpha\otimes\beta\otimes\gamma), \gamma = PD(C). \tag{52}$$

We denote a quantum type cohomology [11, 13] of M by

$$QH^*(M) := H^*(M)\otimes Q[q] \tag{53}$$

where $\mathbb{Q}[q]$ is the ring of Laurent polynomials in q of degree $2N$ with coefficients in the rational numbers $\mathbb{Q}$. By linearly extending the product $*$ on $QH^*(M)$, we have

Theorem 3.1.2. *The quantum type cohomology $QH^*(M)$ of the manifold M is an associative ring under the product $*$.*

Let $(M_1^{2n_1}, g_1, J_1, \omega_1)$ be a symplectic manifold and $(M_2^{2n_2+1}, g_2, \varphi_2, \eta_2, \xi_{2,}, \phi_2)$ be an either almost cosymplectic or contact or C-manifold.

Let the product $(M^{2n+1}, g, \varphi, \eta, \xi, \phi)$ be construct as Theorem 3.2 where $n = n_1 + n_2$. Now we will only consider the free parts of the cohomologies. By the Künneth formula, $H^*(M)\simeq H^*(M_1)\otimes H^*(M_2)$ in particular, $H_2(M)\simeq H_2(M_1\oplus(H_1(M_1)\otimes H_1(M_2))\oplus H_2(M_2))$.

Assume that a two-dimensional class$A = A_1 + A_2 \in H_2(M_1)\oplus H_2(2)\subset H_2(M)$.

Lemma 3.1.3. *Let $(M, g, \varphi, \eta, \xi, \phi)$ be the product $M = M_1 M_2$ constructed as above. For a generic almost cocomplex structure φ on M*

(1) *the moduli space $\mathcal{M}_{0,3}(M; A, \varphi)$ is homeomorphic to the product*

$$\mathcal{M}_{0,3}(M_1, A_1, J_1)\mathcal{M}_{0,3}(M_2, A_2, \varphi), \tag{54}$$

$$\dim\mathcal{M}_{0,3}(M, A, \varphi) = 2[c_1(TM_1)(A_1) + c_1(\mathfrak{H}_2)(A_2)] + 2(n_1 + n_2). \tag{55}$$

Theorem 3.1.4. *For the product $(M, g, \varphi, \eta, \xi, \phi) = (M_1, g_1, J_1, \omega_1)(M_2, g_2, \varphi_2, \eta_2, \xi_2, \phi_2)$, if $A = A_1 + A_2 \in H_2(M_1)\oplus H_2(M_2)\subset H_2(M)$, then the Gromov-Witten type invariants satisfy the following equality*

$$\Phi_{0,3}^{M,A,\varphi} = \Phi_{0,3}^{M_1,A_1,J_1} \cdot \Phi_{0,3}^{M_2,A_2,\varphi_2}. \tag{56}$$

The complex $(n_1 + n_2)$-dimensional vector bundle

$$TM_1 \oplus \mathfrak{H}_2 \to M = M_1 M_2 \tag{57}$$

has the first Chern class $c_1(TM_1 \oplus \mathfrak{H}_2) = c_1(TM_1) + c_1(\mathfrak{H}_2)$.
The minimal Chern numbers N_1 and N_2 are given by $N_1\mathbb{Z} =< c_1(TM_1), H_2(M_1) >$ and

$$N_2\mathbb{Z} =< c_1(\mathfrak{H}_2), H_2(M_2) > . \tag{58}$$

For cohomology classes

$$\alpha = \alpha_1 \otimes \alpha_2 \in H^{k_1}(M_1) \otimes H^{k_2}(M_2) \subset H^k(M), \tag{59}$$

$$\beta = \beta_1 \otimes \beta_2 \in H^{l_1}(M_1) \otimes H^{l_2}(M_2) \subset H^l(M), \tag{60}$$

$k_1 + k_2 = k$, the quantum type product $\alpha * \beta$ is defined by

$$\alpha * \beta = \sum_{\substack{A_1 \in H_2(M_1) \\ A_2 \in H_2(M_2)}} (\alpha_1 * \beta_1)_{A_1} q^{c_1(A_1)/N_1} \otimes (\alpha_2 * \beta_2)_{A_2} q^{c_1(A_2)/N_2} \tag{61}$$

where q_i is a degree $2N_i$ auxiliary variable, $i = 1, 2$, and the cohomology class $(\alpha_i * \beta_i)_{A_i} \in H_{k_i + l_i - 2c_1(A_i)}(M_i)$ is defined by the Gromov-Witten type invariants as follows:

$$\int_{C_i} (\alpha_i * \beta_i)_{A_i} = \Phi_{0,3}^{M_i,A_i,\varphi_i}(\alpha_i \otimes \beta_i \otimes \gamma_i) \tag{62}$$

where $C_i \in H_{k_i + l_i - 2c_1(A_i)}(M_i)$, $\gamma_i = PD(C_i)$ and $\varphi_1 := J_1, i = 1, 2$, respectively.
The quantum type cohomology of M is defined by the tensor product

$$QH^*(M) = H^*(M) \otimes Q[q_1, q_2], \tag{63}$$

where $\mathbb{Q}[q_1, q_2]$ is the ring of Laurent polynomials of variables q_1 and q_2 with coefficients in $\mathbb{Q}$. Extend the product $*$ linearly on the quantum cohomology $QH^*(M)$; similarly, we define the quantum cohomology rings

$$\begin{cases} QH^*(M_1) = H^*(M_1) \otimes Q[q_1], \\ QH^*(M_2) = H^*(M_2) \otimes Q[q_2]. \end{cases} \tag{64}$$

Theorem 3.1.5. *There is a natural ring isomorphism between quantum type cohomology rings constructed as above,*

$$QH^*(M) = QH^*(M_1) \otimes QH^*(M_2). \tag{65}$$

Let (M, g, φ, ϕ) be the product of a compact symplectic manifold $(M_1^{2n_1}, g_1, J_1, \omega_1)$ and an either almost cosymplectic or contact or C-manifold $(M_2^{2n_2+1}, g_2, \varphi_2, \eta_2, \xi_2, \phi_2)$. We choose integral bases, $e_0, e_1, \ldots, e_{k_1}$ of $H^*(M_1)$ and $f_0, f_1, \ldots, f_{k_2}$ of $H^*(M_2)$ such that $e_0 = 1 \in H^0(M_1)$, $f_0 = 1 \in H^0(M_2)$ and each basis element has a pure degree. We introduce a linear polynomial of $k_1 + 1$ variables $t_0, t_1, \ldots, t_{k_1}$, with coefficients in $H^*(M_1)$

$$a_t := t_0 e_0 + t_1 e_1 + \cdots + t_{k_1} e_{k_1}, \tag{66}$$

and a linear polynomial of $k_2 + 1$ variables $s_0, s_1, \cdots, s_{k_2}$ with coefficients in $H^*(M_2)$

$$a_s := s_0 f_0 + s_1 f_1 + \cdots + s_{k_2} f_{k_2}. \tag{67}$$

By choosing the coefficients in $\mathbb{Q}$, the cohomology of M is

$$H^*(M) \cong H^*(M_1) \otimes H^*(M_2). \tag{68}$$

Then, $H^*(M)$ has an integral basis $\{e_i \otimes f_i | i = 0, \ldots, k_1, j = 0, \ldots, k_2\}$. The rational Gromov-Witten type potential of the product (M, ω) is a formal power series in the variables $\{t_i, s_j | i = 0, \ldots, k_1, j = 0, \ldots, k_2\}$ with coefficients in the Novikov ring Λ_ω as follows:

$$\begin{aligned} \Psi_0^M(t,s) &= \sum_A \sum_m \frac{1}{m!} \Phi_{0,m}^{M,A,\varphi}(a_t \otimes a_s, \ldots, a_t \otimes a_s) e^{-\int_A \phi} \\ &= \sum_{A_1} \sum_{m_1} \frac{1}{m_1!} \Phi_{0,m_1}^{M_1,A_1,J_1}(a_t, \ldots, a_t) e^{-\int_{A_1} \omega_1} \cdot \sum_{A_2} \sum_{m_2} \frac{1}{m_2!} \Phi_{0,m_2}^{M_2,A_2,J_2}(a_s, \ldots, a_s) e^{-\int_{A_2} \phi_2} \\ &= \Psi_0^{M_1}(t) \cdot \Psi_0^{M_2}(s). \end{aligned} \tag{69}$$

Theorem 3.1.6. *The rational Gromov-Witten type potential of (M, φ) is the product of the rational Gromov-Witten potentials of M_1 and M_2, that is,*

$$\Psi_0^M(t,s) = \Psi_0^{M_1}(t) \cdot \Psi_0^{M_2}(s). \tag{70}$$

3.2. Floer type cohomology

In this subsection, we assume that our manifold $(M^{2n+1}, g, \varphi, \eta, \xi, \phi)$ is either a almost cosymplectic, contact, or C-manifold.

Let $H_t = H_{t+1} : M \to R$ be a smooth 1-periodic family of Hamiltonian functions. Denoted by $X_t : M \to TM$ the Hamiltonian vector field of H_t.

The vector fields X_t generate a family of Hamiltonian contactomorphisms $\psi_t : M \to M$ satisfying $\frac{d}{dt}\psi_t = X_t \circ \psi_t$ and $\psi_0 = \mathrm{id}$.

Let $a : \mathbb{R}/\mathbb{Z} \to M$ be a contractible loop, then there is a smooth map $u : D \to M$, defined on the unit disk $D = \{z \in C || z| \leq 1\}$, which satisfies $u(e^{2\pi i t}) = a(t)$. Two such maps $u_1, u_2 : D \to M$ are called equivalent if their boundary sum $u_1 \#(-u_2) : S^2 \to M$ is homologus to zero in $H_2(M)$.

http://dx.doi.org/10.5772/65663

Let $\tilde{a} := (a, [u])$ be an equivalence class and denoted by $\widetilde{LM}$ the space of equivalence classes. The space $\widetilde{LM}$ is the universal covering space of the space LM of contractible loops in M whose group of deck transformation is $H_2(M)$.

The symplectic type action functional $a_H : \widetilde{LM} \to R$ is defined by

$$a_H(a, [u]) = -\int_D u^*\phi - \int_0^1 H_t\Big(a(t)\Big)dt, \tag{71}$$

then satisfies $a_H(A\#\tilde{a}) = a_H(\tilde{a}) - \phi(A)$.

Lemma 3.2.1. *Let (M, ϕ) the manifold with a closed fundamental 2-form ϕ and fix a Hamiltonian function $H \in C^\infty(\mathbb{R}/\mathbb{Z}M)$. Let $(a, [u]) \in LM$ and $V \in T_a LM = C^\infty(\mathbb{R}/\mathbb{Z}, a^*TM)$. Then*

$$(da_H)_{(a,[u])}(V) = \int_0^1 \phi\Big(\dot{a} - X_{H_t}(a), V\Big)dt. \tag{72}$$

We denote by $P(\tilde{H}) \subset \widetilde{LM}$ the set of critical points of a_H and by $P(H) \subset LM$ the corresponding set of periodic solutions.

Consider the downward gradient flow lines of a_H with respect to an L^2-norm on LM. The solutions are

$$u : \mathbb{R}^2 \to M, (s, t) \mapsto u(s, t) \tag{73}$$

of the partial differential equation

$$\partial_s(u) + \varphi(u)\Big(\partial_t u - X_t(u)\Big) = 0 \tag{74}$$

with periodicity condition

$$u(s, t + 1) = u(s, t) \tag{75}$$

and limit condition

$$\lim_{s \to -\infty} u(s, t) = a(t), \lim_{s \to +\infty} u(s, t) = b(t), \tag{76}$$

where $a, b \in P(H)$.

Let $\mathcal{M}(\tilde{a}, \tilde{b}) := M(\tilde{a}, \tilde{b}, H, \varphi)$ be the space of all solutions $u(s, t)$ satisfying (74)–(76) with

$$\tilde{a}\#u = \tilde{b}. \tag{77}$$

The solutions are invariant under the action $u(s, t) \mapsto u(s + r, t)$ of the time shift $r \in R$. Equivalent classes of solutions are called Floer connecting orbits.

For a generic Hamiltonian function H, the space $\mathcal{M}(\tilde{a},\tilde{b})$ is a finite dimensional manifold of dimension

$$\dim\mathcal{M}(\tilde{a},\tilde{b}) = \mu(\tilde{a})-\mu(\tilde{b}), \tag{78}$$

where the function $\mu : P(\tilde{H}) \to Z$ is a version of the Maslov index defined by the path of unitary matrices generated by the linealized Hamiltonian flow along $a(t)$ on D .

If $H_t \equiv H$ is a C^2-small Morse function, then a critical point $(a,[u])$ of H_t is a constant map $u(D) = a$ with index $\mathrm{ind}_H(a)$.

If $\mu(\tilde{a})-\mu(\tilde{b}) = 1$, then the space $\mathcal{M}(\tilde{a},\tilde{b})$ is a one-dimensional manifold with $\mathbb{R}$ action by time shift and the quotient $\mathcal{M}(\tilde{a},\tilde{b})/\mathbb{R}$ is a finite set. In fact, $\mu(\tilde{a})\in\pi_1(U(n))\simeq Z$.

If $\mu(\tilde{a})-\mu(\tilde{b}) = 1$, $\tilde{a},\tilde{b}\in P(\tilde{H})$, then we denote

$$\eta(\tilde{a},\tilde{b}) := \#\left(\frac{\mathcal{M}(\tilde{a},\tilde{b})}{\mathbb{R}}\right), \tag{79}$$

where the connection orbits are to be counted with signs determined by a system of coherent orientation s of the moduli space $\mathcal{M}(\tilde{a},\tilde{b})$. These numbers give us a Floer type cochain complex.

Let $FC^*(M,H)$ be the set of functions

$$\xi : P(\tilde{H}) \to R \tag{80}$$

that satisfy the finiteness condition

$$\#\{\tilde{x}\in P(\tilde{H}) | \xi(\tilde{x})\neq 0, a_H(\tilde{x})\leq c\} < \infty \tag{81}$$

for all $c\in\mathbb{R}$.

Now we define a coboundary operator

$$\delta^k : FC^k(M,H) \to FC^{k+1}(M,H), \tag{82}$$

$$(\delta^k\xi)(\tilde{a}) = \sum_{\mu(\tilde{a})=\mu(\tilde{b})+1} \eta(\tilde{a},\tilde{b})\xi(\tilde{b}) \tag{83}$$

where $\xi\in FC^k(M,H)$, $\mu(\tilde{a}) = k+1$ and $\mu(\tilde{b}) = k$.

Lemma 3.2.2. *Let (M,φ) be a semipositive almost contact metric manifold with a closed functional 2-forms. The coboundary operators satisfy $\delta^{k+1}\circ\delta^k = 0$, for all k.*

Definition - Theorem 3.2.3. (1) *For a generic pair (H,φ) on M, the cochain complex (FC^*,δ) defines cohomology groups*

$$FH^*(M, \phi, H, \varphi) := \frac{\text{Ker}\delta}{\text{Im}\delta} \tag{84}$$

which are called the Floer type cohomology groups of the (M, ϕ, H, φ).

(2) *The Floer type cohomology group* $FH^*(M, \phi, H, \varphi)$ *is a module over Novikov ring* Λ_ϕ *and is independent of the generic choices of* H *and* φ.

4. Quantum and Floer type cohomologies

In this section, we assume that our manifold M is a compact either almost cosymplectic or contact or C-manifold. In Section 3.1, we study quantum type cohomology of M and in Section 3.2 Floer type cohomology of M. Consequently, we have:

Theorem 4.1. *Let* $(M, g, \varphi, \eta, \xi, \phi)$ *be a compact semipositive almost contact metric manifold with a closed fundamental 2-form* ϕ. *Then, for every regular pair* (H, φ), *there is an isomorphism between Floer type cohomology and quantum type cohomology*

$$\Phi : FH^*(M, \phi, H, \varphi) \tilde{\rightarrow} QH^*(M, \Lambda_\phi). \tag{85}$$

Proof. Let $h : M \to R$ be a Morse function such that the negative gradient flow of h with respect to the metric $\phi\Big(\cdot, \varphi(\cdot)\Big) + \eta \otimes \eta$ is Morse-Smale and consider the time-independent Hamiltonian

$$H_t := -\varepsilon h, t \in R. \tag{86}$$

If ε is sufficiently small, then the 1-periodic solutions of

$$\dot{a}(t) = X_t(a(t)) \tag{87}$$

are precisely the critical point of h. The index is

$$\mu(a, u_a) = n - \text{ind}_h(a) = \text{ind}_{-h}(a) - n \tag{88}$$

where $u_a : D \to M$ is the constant map $u_a(z) = a$.

The downward gradient flow lines $u : \mathbb{R} \to M$ of h are solutions of the ordinary differential equation

$$\dot{u}(s) = J(u) X_t(u). \tag{89}$$

These solutions determine a coboundary operator

$$\delta : C^*(M, h, \Lambda_\phi) \to C^*(M, h, \Lambda_\phi). \tag{90}$$

This coboundary operator is defined on the same cochain complex as the Floer coboundary δ, and the cochain complex has the same grading for both complex $C^*(M, h, \Lambda_\phi)$, which can be identified with the graded Λ_ϕ module of all functions

$$\xi : \mathrm{Crit}(h)H_2(M) \to R \tag{91}$$

that satisfy the finiteness condition

$$\#\{(a, A)|\xi(a, A)\neq 0, \phi(A)\geq c\} < \infty \tag{92}$$

for all $c\in R$. The Λ_ϕ-module structure is given by

$$(\nu * \xi)(a, A) = \sum_B \nu(B)\xi(a, A + B), \tag{93}$$

the grading is $\deg(a, A) = \mathrm{ind}_h(a) - 2c_1(A)$, and the coboundary operator δ is defined by

$$(\delta\xi)(a, A) = \sum_b n_h(a, b)\xi(b, A), (a, A)\in \mathrm{Crit}(h)H_2(M), \tag{94}$$

where $n_h(a, b)$ is the number of connecting orbits from a to b of shift equivalence classes of solutions of

$$\{ \dot{u}(s) + \nabla u(s) = 0, \lim_{s\to-\infty} u(s) = a, \lim_{s\to+\infty} u(s) = b, \tag{95}$$

counted with appropriate signs.

Here we assume that the gradient flow of h is Morse-Smale and so the number of connecting orbits is finite when $\mathrm{ind}_h(a) - \mathrm{ind}_h(b) = 1$. Then the coboundary operator δ is a Λ_ϕ-module homomorphism of degree one and satisfies $\delta\circ\delta = 0$. Its cohomology is canonically isomorphic to the quantum type cohomology of M with coefficients in Λ_ϕ.

For each element $\tilde{a}\in P(\tilde{H})$ we denote $\mathcal{M}(\tilde{a}, H, \varphi)$ by the space of perturbed φ-cohomomorphic maps $u : \mathbb{C} \to M$ such that $u(re^{2\pi i t})$ converges to a periodic solution $a(t)$ of the Hamiltonian system H_t as $r \to \infty$. The space $\mathcal{M}(\tilde{a}, H, \varphi)$ has dimension $n - \mu(\tilde{a})$. Now fix a Morse function $h : M \to R$ such that the downward gradient flow $u : \mathbb{R} \to M$ satisfying (95) is Morse-Smale. For a critical point $b\in \mathrm{Crit}(h)$ the unstable manifold $W^u(b, h)$ of b has dimension $\mathrm{ind}_h(b)$ and codimension $2n - \mathrm{ind}_h(b)$ in the distribution D.

The submanifold $\mathcal{M}(b, \tilde{a})$ of all $u\in M(\tilde{a}, H, \varphi)$ with $u(0)\in W^u(b)$ has dimension

$$\dim\mathcal{M}(b, \tilde{a}) = \mathrm{ind}_h(b) - \mu(\tilde{a}) - n. \tag{96}$$

If $\mathrm{ind}_h(b) = \mu(\tilde{a}) + n$, then $\mathcal{M}(b, \tilde{a})$ is 0zero-dimensional and hence the numbers $n(b, \tilde{a})$ of its elements can be used to construct the chain map defined by

$$\Phi : FC^*(M, H) \to C^*(M, h, \Lambda_\phi) \tag{97}$$

$$(\Phi\xi)(b, A)\Lambda \sum_{\mathrm{ind}_h(b) = \mu(\tilde{a}) + n} n(b, \tilde{a})\xi(A\#\tilde{a}) \tag{98}$$

which is a Λ_ϕ-module homomorphism and raises the degree by n. The chain map Φ induces a homomorphism on cohomology

$$\Phi : FH^*(M,\Lambda_\phi) \to H^*(M,h,\Lambda_\phi) = \frac{\mathrm{Ker}\delta}{\mathrm{Im}\delta} \simeq QH^*(M,\Lambda_\phi). \tag{99}$$

Similarly, we can construct a chain map,

$$\Psi : C^*(M,h,\Lambda_\phi) \to FC^*(M,H) \tag{100}$$

$$(\Psi\xi)(\tilde{a}) := \sum_{\mu(\tilde{a})+n=\mathrm{ind}_h(b)-2c_1(A)} n((-A)\#\tilde{a},b)\xi(b,A). \tag{101}$$

Then $\Phi\circ\Psi$ and $\Psi\circ\Phi$ are chain homotopic to the identity. Thus we have an isomorphism Φ.

We have studied the Gromov-Witten invariants on symplectic manifolds (M,ω,J) using the theory of J-holomorphic curves, and the Gromov-Witten type invariants on almost contact metric manifolds $(N,g,\varphi,\eta,\xi,\phi)$ with a closed fundamental 2-form ϕ using the theory of φ-coholomorphic curves. We also have some relations between them. We can apply the theories to many cases.

Examples 4.2.

1. The product of a symplectic manifold and a unit circle.
2. The circle bundles over symplectic manifolds.
3. The almost cosymplectic fibrations over symplectic manifolds.
4. The preimage of a regular value of a Morse function on a Kähler manifold.
5. The product of two cosymplectic manifolds is Kähler.
6. The symplectic fibrations over almost cosymplectic manifolds.
7. The number of a contactomorphism is greater than or equal to the sum of the Betti numbers of an almost contact metric manifold with a closed fundamental 2-form.

Examples 4.3. Let N be a quintic hypersurface in $\mathbb{C}P^4$ which is called a Calabi-Yau threefold. Then N is symply connected, $c_1(TN)=0$ and its Betti numbers $b_0=b_6=1,\ b_1=b_5=0$, $b_2=b_4=1$ and $b_3=204$.

Let A be the standard generator in $H_2(N)$ and $h\in H^2(N)$ such that $h(A)=1$. The moduli space $\mathcal{M}_{0,3}(N,A)$ has the dimension zero. The Gromov-Witten invariant $\Phi_{0,3}^{N,A}(a_1,a_2,a_3)$ is nonzero only when $\deg(a_i)=2,\ i=1,2,3$. In fact, $\Phi_{0,3}^{N,A}(h,h,h)=5$ [4, 5]. The quantum cohomology of N is $QH^*(N)=H^*(N)\otimes\Lambda$ where Λ is the universal Novikov ring [5].

Let (N,g_1,ω_1,J_1) be the standard Kähler structure on N and $(S^1,g_2,\varphi_2=0,\eta_2=d\theta$, $\xi_2=\frac{d}{d\theta},\phi_2=0)$ the standard contact structure on S^1. Then the product $M=NS^1$ has a canonical cosymplectic structure $(M,g,\varphi,\eta,\xi,\phi)$ as in Section 3. The quantum type cohomology of M is

$$QH^*(M) = QH^*(N) \otimes QH^*(S^1) \tag{102}$$

Let $\psi_1 : N \to N$ be a Hamiltonian symplectomorphism with nondegenerate critical points. Then $\#Fix(\psi_1) \geq \sum_{i=0}^{6} b_i(N) = 208$.

Let $\psi_2 : M \to M$ be a Hamiltonian contactomorphism with nondegenerate critical points. Then $\#Fix(\psi_2) \geq \sum_{i=0}^{7} b_i(M) = 416$.

Author details

Yong Seung Cho

Address all correspondence to: yescho@ewha.ac.kr

Department of Mathematics, Ewha Womans University, Seoul, Republic of Korea

References

[1] D. E. Blair and S. I. Goldberg. Topology of almost contact manifolds. J. Differ. Geom. 1967;**1**:347–354.

[2] Y. S. Cho. Finite group actions on the moduli space of self-duel connections. Trans. Am. Math. Soc. 1991;**323**(1):233–261.

[3] Y. S. Cho. Hurwitz number of triple ramified covers. J. Geom. Phys. 2008;**56**(4):542–555.

[4] Y. S. Cho. Quantum cohomologies of symmetric products. Int. J. Geom. Methods Mod. Phys. 2012;**9**(5):1250045.

[5] Y. S. Cho. Quantum type cohomologies on contact manifolds. Int. J. Geom. Methods Mod. Phys. 2013;**10**(5):1350012.

[6] Y. S. Cho. Stabilization of 4-Manifolds. Sci. China Math. 2014;**57**(9):1835–1840.

[7] Y. S. Cho. Floer type cohomology on cosymplectic manifolds. Int. J. Geom. Methods Mod. Phys. 2015;**12**(7):1550082.

[8] Y. S. Cho and S. T. Hong. Dynamics of stringy congruence in the early universe. Phys. Rev. D. 2011;**83**:104040.

[9] A. Floer. Symplectic fixed points and holomorphic spheres. Commun. Math. Phys. 1989;**120**(4):575–611.

http://dx.doi.org/10.5772/65663

[10] D. Janssens and J. Vanhecke. Almost contact structures and curvature tensors. Kodai Math. J. 1991;**4**:1–27.

[11] M. Kontsevich and Y. Manin. Gromov-Witten classes, quantum cohomology and enumerative geometry. Commun. Math. Phys. 1994;**164**:525–562.

[12] D. McDuff and D. Salamon. *J*-holomorphic curves and quantum cohomology. University Lecture Series, 6. American Mathematical Society, Providence, RI. 1994.

[13] D. McDuff and D. Salamon. Introduction to symplectic topology. Oxford University Press, Oxford. 2005.

Section 4

Manifolds and Applications

Chapter 6

Head Pose Estimation via Manifold Learning

Chao Wang, Yuanhao Guo and Xubo Song

Additional information is available at the end of the chapter

http://dx.doi.org/10.5772/65903

Abstract

For the last decades, manifold learning has shown its advantage of efficient non-linear dimensionality reduction in data analysis. Based on the assumption that informative and discriminative representation of the data lies on a low-dimensional smooth manifold which implicitly embedded in the original high-dimensional space, manifold learning aims to learn the low-dimensional representation following some geometrical protocols, such as preserving piecewise local structure of the original data. Manifold learning also plays an important role in the applications of computer vision, i.e., face image analysis. According to the observations that many face-related research is benefitted by the head pose estimation, and the continuous variation of head pose can be modelled and interpreted as a low-dimensional smooth manifold, we will focus on the head pose estimation via manifold learning in this chapter. Generally, head pose is hard to directly explore from the high-dimensional space interpreted as face images, which is, however, can be efficiently represented in low-dimensional manifold. Therefore, in this chapter, classical manifold learning algorithms are introduced and the corresponding application on head pose estimation are elaborated. Several extensions of manifold learning algorithms which are developed especially for head pose estimation are also discussed and compared.

Keywords: manifold learning, head pose estimation, nonlinear feature reduction, supervised manifold learning, local linearity, global geometry

1. Introduction

Manifold learning becomes well known due to its property to learn the representative geometry in low-dimensional embedding, with which data analysis and visualization are significantly benefitted. From the observation of some nonlinear data, a low-dimensional smooth manifold (differentiable manifold) is embedded in the original high-dimensional space, which is implicit if we only consider the metrics of the original space. Manifold learning algorithm purpose is to learn such embedding according to some protocols, e.g., local linearity and global structure preserving. For the remainder of this chapter, the term manifold is used to refer as the smooth manifolds (differentiable manifolds) for convenience. As a complex high-dimensional data, face image analysis is a difficult topic in the field of computer vision due to the complicated facial appearance variations, among which the head pose challenges many face-related applications. Accurate head pose estimation is advantageous to face alignment and recognition, because frontal- or near-frontal faces are easier to handle compared with other poses. It has been found that facial appearance is lying on a manifold embedded form in the original high-dimensional space represented as face images. Correspondingly, the head pose can also be represented as a low-dimensional embedding, which is more representative and discriminative to model the variation. Therefore, the head pose estimation can be implemented by the manifold learning.

In principle, head pose refers to the view of the face to the imaging system, i.e., the camera center. 3D head transformation involves 6 degrees of freedom (DOF), which can be interpreted as the 3D translations $(t_x, t_y, t_z)^T$ and rotations $(\alpha, \theta, \gamma)^T$ from the head to the camera center. Among the six variables, the 3D rotations that are formally represented by pitch, roll, and yaw are taken as the head pose [1]. A schematic demonstration taken from reference [1] is shown in **Figure 1**. This definition reduces the head pose to 3 DOF which are sufficient to model most of the in- and out-plane rotation of the head. It can be found that the pitch and yaw generate more self-occlusions and roll can be easily corrected by the position of eyes. So, in this chapter, the 2 DOF including yaw and pitch are considered. Usually, the head poses of yaw and pitch lead to the problem of self-occlusion, which subsequently results in the loss of informative features, e.g., facial texture and shapes. By comparing, frontal- or near-frontal faces are relatively easier to deal with. Examples of various head poses [2] are given in **Figure 2**. The task of the head pose estimation is actually to determine the yaw and pitch of an unknown face image or search the frontal face in a database. Once the head pose is obtained, further applications such as face alignment and recognition will be benefitted in efficiency and accuracy. Therefore, modern development of face-related research prefers to estimate the head pose and correct the head orientation by positioning the face to near frontal or warping the faces in various poses to a frontal face template [3].

Basically, head pose estimation methods broadly fall into several categories. Template-based methods treat the head pose estimation as a verification (or classification) problem. The testing face is projected to the data set labeled with known poses, the one from which the most significant similarity measured by various metrics is retrieved for the testing pose [4, 5]. Furthermore, pose detectors can be learned to simultaneously localize the face and recognize the pose [6]. Regression-based methods estimate a linear or nonlinear function with the original faces or extracted facial features as input variables and discrete or continuous poses as output [2]. Deformable models learn flexible facial modes [7–9]. By manipulating a set of parameters which specifies the pose, specific face example can be generated, which will be used to match the testing face. With the development of manifold learning [10–13], more promising results of head pose estimation are achieved. The essence of such methods is based on the assumption that the discriminative modes for head pose lie on low-dimensional manifolds embedded in high-dimensional space, i.e., the original color space or other low level

http://dx.doi.org/10.5772/65903

feature space [14]. The low-dimensional representation of the head pose images can be learned by unsupervised or supervised manifold learning.

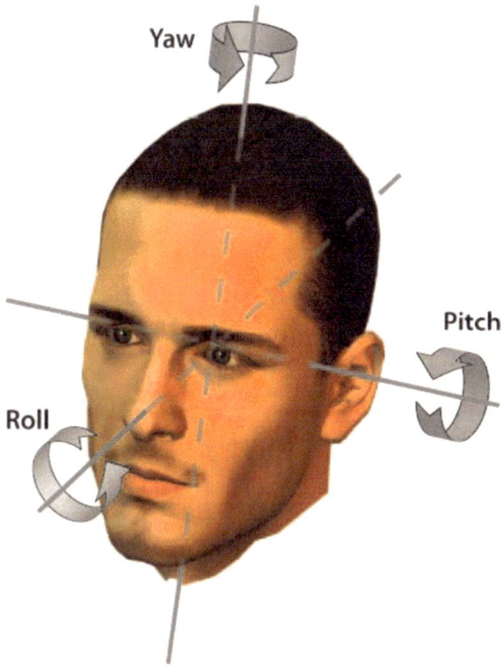

Figure 1. The 3 DOF of head pose proposed in reference [1]. The roll does not introduce any self-occlusion of the face, which can be easily corrected. Compared with the pitch, the yaw produces more serious self-occlusion problem.

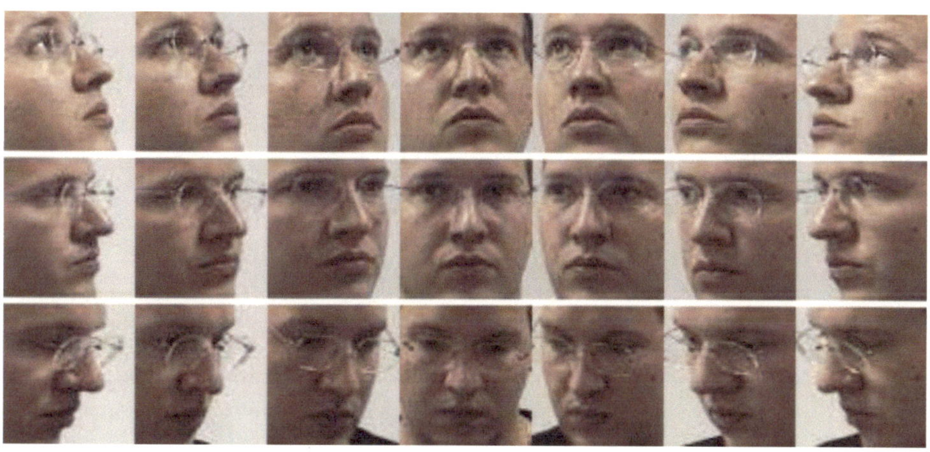

Figure 2. Examples of various head poses. The images are cropped, centered, and resized to 64 × 64 pixels from the originals. One individual is selected and shown in different yaw and pitch. From left to right represents the variation in yaw: −90, −60, −30, 0, 30, 60, and 90°. From top to bottom represents the variation in pitch: 30, 0, and −30°. One can find that the effects of self-occlusion occur with an increasing yaw and pitch. The frontal faces (center image) show a full overview of the face.

In contrast, the template-based methods have the problem of serious dependence on training data. If similar poses to the query pose do not exist in the training set, the estimated result would be biased. The regression-based methods often require to use complicated regression models, for example, a high-order polynomial. However, complicated nonlinear function would cause the problem of overfitting, which will result in poor generalization of the model. The deformable models require the localization of dense facial features, such as landmarks of facial components, which are seriously influenced by the head pose. The manifold learning-based methods are somehow limited by some problems, such as identity and noise sensitivity; however, simple efforts can be made to efficiently improve the performance [15]. More importantly, the manifold learning-based methods show promising performance of generalization. And the head pose can be easily modeled and better visualized with low-dimensional features. More details will be given in following sections.

According to the previous analysis, the main focus of this chapter will be on the manifold learning based on head pose estimation. The main notations used in this chapter are listed and interrupted in Section 2. In Section 3, classical manifold learning algorithms will be elaborated. In Section 4, adaptions and extensions of manifold learning algorithms, which are more suitable for head pose estimation, are discussed. Section 4 summaries the work, and some available resources of manifold learning are given.

2. Notations

$\boldsymbol{x}_i = (x_{i1}, x_{i2}, \dots, x_{iD})^T$ represents the ith data point in the original D-dimensional space. $\boldsymbol{x}$ without subscript is used to represent an arbitrary data point.

$\mathbf{X} = (\boldsymbol{x}_1, \boldsymbol{x}_2, \dots, \boldsymbol{x}_M)$ represents the M-data points collection.

$N(i)$ represents the set of K-nearest neighbor of the ith data point.

$\boldsymbol{y}_i = (y_{i1}, y_{i2}, \dots, y_{id})^T$ represents the d-dimensional representation of the ith data point after dimensionality reduction. Similarly, $\boldsymbol{y}$ without subscript is used to represent an arbitrary data point.

$\mathbf{Y} = (\boldsymbol{y}_1, \boldsymbol{y}_2, \dots, \boldsymbol{y}_M)$ denotes the data collection in the low-dimensional space.

$\mathbf{C} = \mathbf{X}\mathbf{X}^T$ is the covariance matrix for the centered data, where $\frac{1}{M}\sum_{i=1}^{M}\boldsymbol{x}_i = \mathbf{0}$. Usually, this can be enabled by mean subtraction: $\boldsymbol{x} = \boldsymbol{x} - \boldsymbol{\mu}$, and $\boldsymbol{\mu} = \frac{1}{M}\sum_{i=1}^{M}\boldsymbol{x}_i$.

$\mathbf{W} = \{w_{ij}\}$ is the weight matrix to model the graph for pairwise face images, which will be specified by different metrics, e.g. $w_{ij} = 1$ if the jth data point is one of the K nearest neighbors of the ith data point, and $w_{ij} = 0$ otherwise.

$\mathbf{D} = \{d_{ij}\}$ is the distance matrix measuring the pairwise distances among the data points, which can be Euclidean or other distance metrics.

$\Lambda = diag(\lambda_1, \lambda_2, \dots, \lambda_D)$ is a diagonal matrix whose elements are the D eigenvalues (ranked decreasingly $\lambda_1 \geq \lambda_2 \geq \dots \geq \lambda_D \geq 0$ decomposed from an $D \times D$ matrix.

$\mathbf{V} = (v_1, v_2, \ldots, v_d, \ldots, v_{D-d+1}, \ldots, v_D)$ is the projection matrix, which is consisted of the eigenvectors corresponding to the ranked eigenvalues. $(v_1, v_2, \ldots, v_d)$ are the top d eigenvectors, and $(v_{D-d+1}, \ldots, v_D)$ are the bottom d eigenvectors.

$\mathbf{I}$ denotes the identity matrix.

3. Characteristics of manifold learning algorithms

Given a set of data points, for example, face images, it is difficult to directly estimate or extract the most significant modes from such high-dimensional representation of the data. If the distribution of data in the original feature space can be linearly structured, the classical principal component analysis (PCA) will be able to estimate the discriminative modes and then reduce the feature dimensions. An example of such a type of data is shown in **Figure 3**. However, if the data distribution of the original data is nonlinear, for example, the famous "swiss roll" shown in **Figure 4(a)**, which is a smooth, continuous but nonlinear surface embedded in the 3D space, the structure interpreted as Euclidean distance is less preferable to represent the distribution of the data. Taking the two circled points sampled from the manifold shown in **Figure 4(b)**, for instance, their Euclidean distance is close, while this is not guaranteed if the 3D structure is considered. The embedded structure can be explored with the help of nonlinear dimensionality reduction, such as manifold learning algorithms. The learned low-dimensional representation can approximately model the real distance of the sampled data points as shown in **Figure 4(c)**.

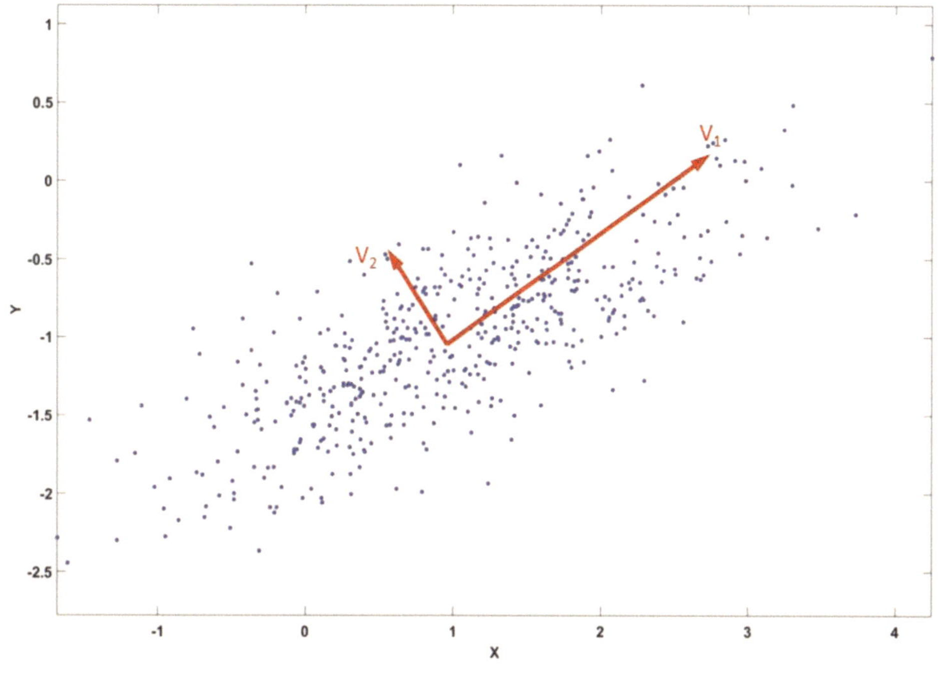

Figure 3. A data set sampled from a multivariate Gaussian distribution. The most significant modes indicated by the red orthogonal axis can be learned by PCA, which preserve the largest variations in the original data.

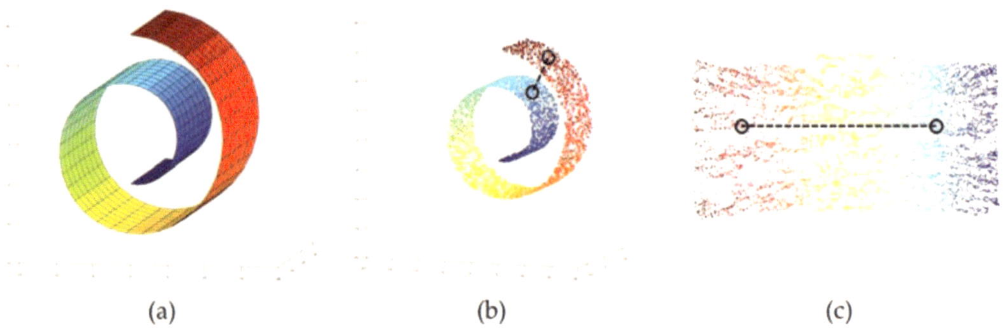

Figure 4. An example of the data set including a potential "swiss roll" structure. The figures are produced based on the code from [16]. (a) The original 3D surface. (b) The data points sampled from (a). The Euclidean distance indicated by dash line between the circled points cannot represent the distance lying on the potential structure. (c) The distance measured in the learned low-dimensional can more accurately model the data.

In this section, in order to reveal the essence of manifold learning, the PCA is initially detailed. Other classical manifold learning algorithms will be elaborated in the following.

3.1. Principal component analysis (PCA)

PCA is one of the most popular unsupervised linear dimensionality reduction algorithms. The intrinsic feature of PCA is to estimate a linear space whose basis will be able to preserve the maximum variations in the original data. Mathematically, the low-dimensional data can be obtained by a linear transformation from the original data as denoted in Eq. (1).

$$y = \mathbf{V}^T x \tag{1}$$

where x is the centered data point. The entry of the projection matrix $\mathbf{V}$ is the column vector that represents the principal components in the projection space. Let us take one of the principal components for instance. The objective is to preserve the maximum variations in the transformed data.

$$\max_{\|v\|=\mathbf{1}} Var(v^T\mathbf{X}) = \max_{v} \frac{1}{M-1}\|v^T\mathbf{X}\|^2 = \max_{v} \frac{1}{M-1}(v^T\mathbf{X})(\mathbf{X}^T v) = \max_{v}(v^T\mathbf{C}v) \tag{2}$$

$\frac{1}{M-1}$ makes Eq. (2) an unbiased estimation, which can be replaced by M if it is sufficiently large. Eq. (2) is a form of Rayleigh's quotient, which can be maximized by eigenvector decomposition of covariance matrix $\mathbf{C}$. The d eigenvectors corresponding to the top d positive eigenvalues are taken to construct the low-dimensional space $V_{D \times d}$ to which the original data are projected. Actually, Eq. (2) can be converted to

$$x = \mathbf{V}y \tag{3}$$

which means that the original data can be linearly represented as a combination of the principal components.

http://dx.doi.org/10.5772/65903

Taking the head pose images of one identity shown in **Figure 2**, for example, the PCA is applied on the vectorized images. **Figure 5** visualizes the low-dimensional representation of the face images in the first 3D dimensions. One can find obvious transitions for pitch and yaw along a 3D shape of valley. The three principal components are visualized in **Figure 6**, from which one face image is decomposed into a weighted accumulation of variations in the mean face. The first and third eigenfaces (principal component) clearly show the variation in yaw. Therefore, PCA can model the head poses as some of the discriminative principal components.

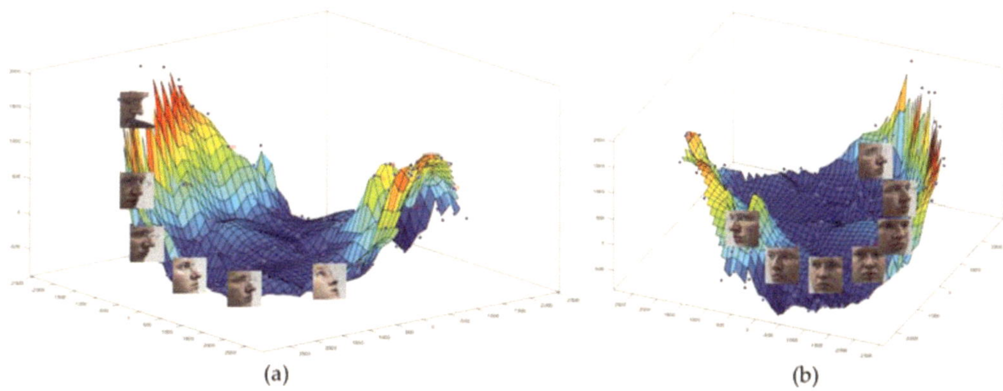

Figure 5. Visualization of the low-dimensional features obtained by PCA. A surface of valley can be found. Blue dots show the face images sampled from the surface. Some face images are selected and shown. (a) The variation in pitch is shown with the yaw of -90° in a specific view of the surface. (b) The variation in yaw is shown with the pitch of 0° in another view of the surface.

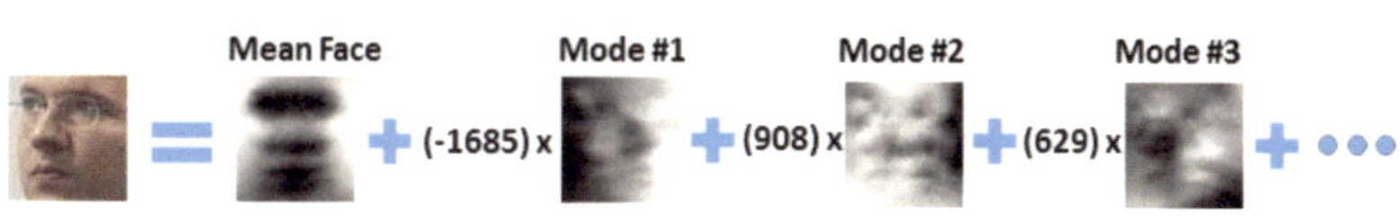

Figure 6. Representation of one face image by the mean face and first three eigenfaces obtained from PCA.

3.2. Locally linear embedding (LLE)

From the observation of the data shown in **Figure 4**, the smooth manifold is globally nonlinear but can be seen as linear from a local neighborhood. On the basis of this observation, the LLE attempts to represent each of the data by a weighted linear combination of a number of neighbors [11]. The weight matrix **W** can be obtained by the following objective function.

$$\min_{\|w_i\|=1} \sum_{i=1}^{M} \left\| x_i - \sum_{j \in N(i)} w_{ij} x_j \right\|^2 \tag{4}$$

where w_i denotes the ith row vector of matrix **W**. Eq. (4) shows that the LLE aims to minimize the total reconstruction error for the data from the corresponding nearest neighbors. Specifically, **W** is a sparse matrix which assigns optimal weights for neighboring data points and zeros for nonneighboring data points. As a result, both the global nonlinearity and local linearity are included in one identical form. A close form of the weights matrix $\mathbf{W} = \{w_{ij}\}$ can be efficiently computed.

$$w_{ij} = \frac{\sum_{n=1}^{K} c_{jn}^{-1}}{\sum_{m=1}^{K}\sum_{n=1}^{K} c_{mn}^{-1}} \tag{5}$$

where K is the number of nearest neighbors tuned for specific problem; $\mathbf{C} = \{c_{mn}\}$ is the neighborhood correlation matrix that is specified for each data: $c_{mn} = \xi_m^T \xi_n$ where ξ_m and ξ_n are the mth and nth neighbors of the ith data point.

Moreover, the weight matrix $\mathbf{W}$ is locally invariant to linear transformation, i.e., translation, rotation, and scaling. Therefore, it is reasonable to propose that low-dimensional representation of the data can also preserve the local geometry as featured in the original space with the same weight matrix $\mathbf{W}$. The next step is then to estimate such low-dimensional (d-dimension) representation $\mathbf{Y}$ by the following equation.

$$\min_{\mathbf{Y}} \sum_{i=1}^{M} \left\| \mathbf{y}_i - \sum_{j \in N(i)} w_{ij} \mathbf{y}_j \right\|^2 \tag{6}$$

Eq. (6) can be rewritten as

$$\min_{\mathbf{Y}} \mathbf{Y}\mathbf{M}\mathbf{Y}^T \tag{7}$$

where $\mathbf{M} = (\mathbf{I}-\mathbf{W})(\mathbf{I}-\mathbf{W})^T$. Two constraints are made to center the data and avoid degenerate solutions: $\sum_{i=1}^{M} \mathbf{y}_i = \mathbf{0}$ and $\frac{1}{M}\sum_{i=1}^{M} \mathbf{y}_i \mathbf{y}_i^T = \mathbf{I}_{d\times d}$. Then $\mathbf{Y}$ can be obtained, of which the columns correspond to the bottom d eigenvectors of $\mathbf{M}$.

The same data set used in the last experiment is processed with LLE and shown with the first three dimensions in **Figure 7**. The variation of the head pose in yaw with different pitch is obviously shown. The transition from a pose to another pose tends to be continuous and easy to locate. The learned manifold is smoother and more discriminative than PCA.

3.3. Isomap

Isomap [10] is an abbreviation of isometric feature mapping [17], which is an extension of the classical algorithm of multidimensional scaling (MDS) [18]. From the previous section, one can learn that the LLE represents the nonlinearity of the original data by preserving the local geometrical linearity. In contrast, the algorithm of Isomap proposed a global solution by constructing a graph for all pairwise data. This idea ensures the global optimum.

Specifically, Isomap firstly constructs a graph that can be represented as $G = (V,E)$ with V as the vertices (the data points) and E as the edges (the adjacent connections). The adjacent vertices will be connected with edges according to particular metrics. Taking the ith and jth data points

http://dx.doi.org/10.5772/65903

for instance, an edge will be added between them if $\|\boldsymbol{x}_i - \boldsymbol{x}_j\| \leq \epsilon$ (ϵ-neighbor) or $\boldsymbol{x}_i$ is the K-nearest neighbor of $\boldsymbol{x}_j$ and vice versa (K-nearest neighbor). The graph is then assigned with distances among all pairwise vertices according to the edges. The distance between the vertices associated with edges will be $d_{ij} = \|\boldsymbol{x}_i - \boldsymbol{x}_j\|$ (Euclidean distance). For the other pairwise vertices, the geodesic distances are considered, which can be simply computed as the shortest path (Floyd's algorithm). This weighted graph is capable of modeling the isometric distances, which can preserve the global geometry in the learned manifold.

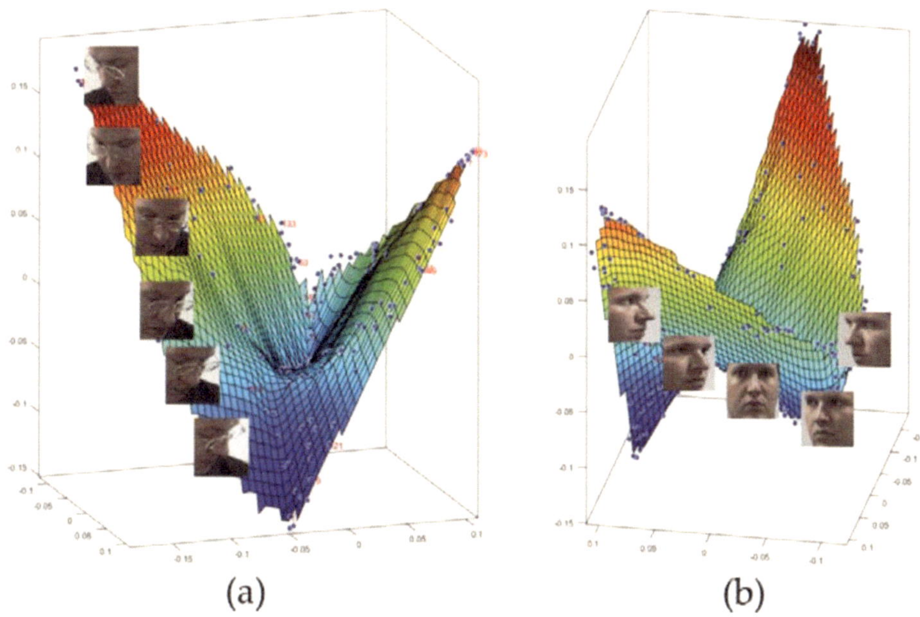

Figure 7. Visualization of the low-dimensional features obtained by LLE. A surface of "wings" is observed. (a) The variation in yaw with the pitch of -30° is found along the edge of one "wing" of the 3D surface. (b) Another variation in yaw with pitch of 0° is found along the ridge of the 3D surface.

From the distance matrix $\mathbf{D} = \{d_{ij}\}$, an objective function is defined as

$$\min_{\mathbf{Y}} \left\| \tau(\mathbf{D}_{\mathbf{X}}) - \tau(\mathbf{D}_{\mathbf{Y}}) \right\|^2 \tag{8}$$

where $\tau(\mathbf{D}_{\mathbf{X}})$ and $\tau(\mathbf{D}_{\mathbf{Y}})$ denote the distances conversion to inner products for the original data and the low-dimensional data, respectively. The operator τ is defined as $\tau(\mathbf{D}) = -\mathbf{HSH}/2$; $\mathbf{S} = \mathbf{D}^2$ is the squared distance matrix; $\mathbf{H} = \mathbf{I} - \frac{1}{M}\mathbf{1}\mathbf{1}^T$ is the centering matrix. This design of the objective function aims to preserve the global structure represented as a graph associated with geodesic distance. The low-dimensional embedding should be featured with similar global geometry with the original data. The optimization of the objective function can be solved by MDS. The d-dimensional representation $\mathbf{Y}$ is obtained by decomposing the matrix of $\tau(\mathbf{D}_{\mathbf{X}})$ and preserving the top d eigenvectors.

Figure 8 shows the results of the low-dimensional head poses obtained from Isomap. An interesting shape of "bowl" of the embedding surface obtained for the head pose images.

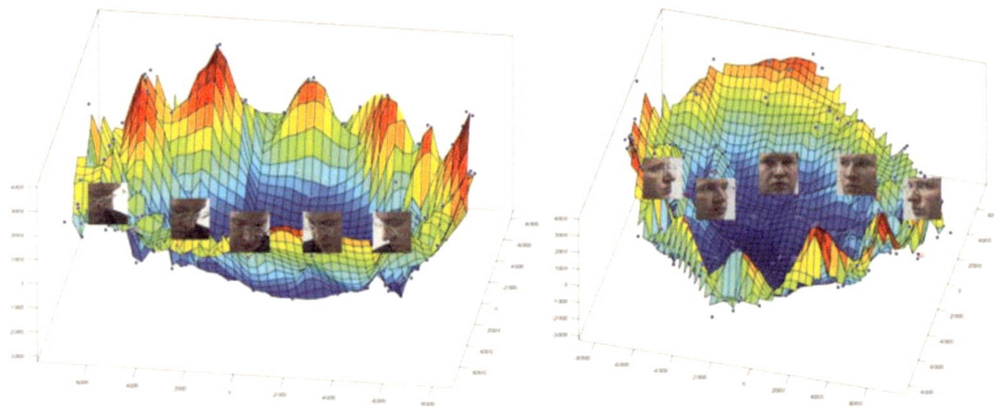

(a) (b)

Figure 8. Visualization of the low-dimensional features obtained by Isomap. (a) The variation in yaw with the pitch of -30° is found along the edge of the shape. (b) Another variation in yaw with pitch of 0° is found along the geodesic path in the middle of the shape. The interesting thing is the frontal face locates approximated at the center.

3.4. Laplacian eigenmaps (LE)

Compared to Isomap, the idea of graph representation of the data is also taken by the algorithm of LE. However, the difference is the later attempts to construct a weighted graph (other than distance graph) for the data, which is then represented as a Laplacian [12].

The first step of LE is to construct an adjacent graph whose vertices are the data points and edges are the adjacent connections for neighbors. A pair of points $\boldsymbol{x}_i$ and $\boldsymbol{x}_j$ are ϵ-neighbors and will be connected with edge if $\|\boldsymbol{x}_i - \boldsymbol{x}_j\| \le \varepsilon$. The other criterion to connect or disconnect the pair of points is to find if they are K-nearest neighbors for each other. The second step is to choose appropriate weights for the graph. There are two options: the heat kernel defines the weight as $w_{ij} = e^{-\frac{\|\boldsymbol{x}_i-\boldsymbol{x}_j\|^2}{t}}$ if the two points of $\boldsymbol{x}_i$ and $\boldsymbol{x}_j$ are connected and zero otherwise. The other option is straightforwardly setting w_{ij} =1 for connected edges and zero otherwise.

The third step is to minimize an objective function

$$\min_{y:y\mathbf{A}y^T=1} \sum_{i,j} (\boldsymbol{y}_i - \boldsymbol{y}_j)^2 w_{ij} \tag{9}$$

where the diagonal matrix of $\mathbf{A} = diag\{a_{ii}\}$ is computed by column sums of $\mathbf{W}$: $a_{ii} = \sum_j w_{ji}$. From the definition of the objective function, the goal of LE is to preserve the weights for the mapped data from the original data. If the pair of data is close or apart from each other in the original space, they should be also kept close or apart in the embedding. The weight matrix strongly punishes the "connection" for apart data points. Next, the objective function can be derived to

$$\mathbf{L}y = \lambda \mathbf{A}y \tag{10}$$

where $\mathbf{L} = \mathbf{A} - \mathbf{W}$ is the Laplacian matrix for the weighted graph. Such form is a generalized eigenvector decomposition. And the matrix $\mathbf{Y}$ whose columns are the bottom d eigenvectors decomposed from Eq. (10) is the d-dimensional representation of the data. To be compared, the LE is less sensitive to outlier and noise due to its property of local preservation. The weights for nonedges are set to be zeros, which diminish the problem of short circuiting.

As shown in **Figure 9**, the embedding surface with the shape of parabolais generated by LE, which is similar to the results obtained by LLE. But the latter produces smoother and more symmetric shape of the surface. The variation in yaw from left to right is shown symmetrically, and the frontal face approximately locates on the vertex of bottom.

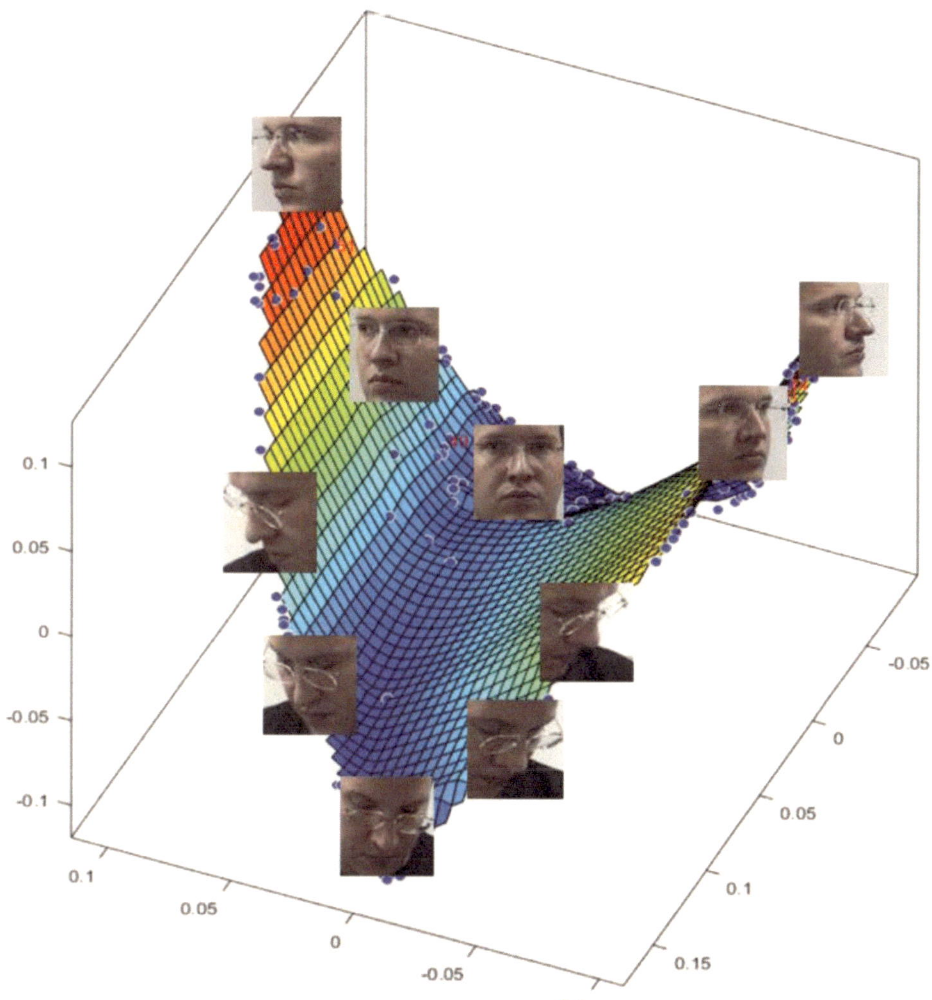

Figure 9. Visualization of the low-dimensional features obtained by LE.

3.5. Laplacian preserving projections (LPP)

The previously introduced algorithms do not clarify how an unseen data is projected to the low-dimensional space. To solve this problem, LPP reformulates the LE by representing the dimensionality reduction as a linear projection from the original to the low-dimensional data. The first two steps of LPP are exactly the same as LE, which construct the adjacent graph and compute the weights for each connection. The most significant difference is the LPP representing the dimensionality reduction from the original to the low-dimensional space as a projection $\boldsymbol{y} = \mathbf{V}^T\boldsymbol{x}$. The problem is converted to the one which aims to find a projection space instead of directly compute the low-dimensional features. The generalized eigenvector decomposition defined in Eq. (10) is then reformulated as follows:

$$\mathbf{XLX}^T v = \lambda \mathbf{XAX}^T v \tag{11}$$

The bottom d eigenvectors decomposed from Eq. (11) construct the projection matrix $\mathbf{V}_{D\times d} = \{v_i\}$. Any data from the original space can be dimensionally reduced through $\boldsymbol{y} = \mathbf{V}^T\boldsymbol{x}$.

More improved nonlinear manifold learning algorithms are developed [13, 19], but in this section, the main idea of how to derive the low-dimensional representation of the head poses is the core. Details of the advanced versions of the manifold learning algorithms can be explored in the original references.

4. Head pose estimation via manifold learning

The manifold learning methods can successfully model the head pose variations in both yaw and pitch as discussed in the previous sections. However, there are still several difficulties to state. The introduction of noise, for example, identity, and illumination variations will affect the performance of those methods on the head pose estimation. Another point is that they do not infer how the low-dimensional representation of an unseen head pose image is obtained (except LPP) and how the pose is estimated. In this section, more sophisticated methods are introduced to solve these problems based on the original or extended manifold learning algorithms.

4.1. PCA-based head pose estimation

In Ref. [20], the PCA has been turned to be robust to invariance of identity. Another important conclusion is that the angle of 10° is found to be the lower bound to be discriminative. For the data set constructed following this finding, the PCA would produce promising results for head pose estimation.

A kernel machine-based method is proposed using the kernel PCA (KPCA) and kernel support vector classifier (KSVC) [21]. The KPCA is an extension of the classical PCA. Let $\phi(\boldsymbol{x}) : R^D \rightarrow R^{D'}$ be a kernel that maps the original dimensional into a higher dimensional, which makes the nonlinearly separable data linearly separable in the higher dimensional space. Correspondingly, the covariance matrix **C** is replaced by $\overline{\mathbf{C}} = \Phi\Phi^T$, of which the bold Φ is the kernel represented data points set. The projection matrix $\overline{\mathbf{V}}$ can be similarly obtained through eigenvector decomposition of matrix $\overline{\mathbf{C}}$. After the feature dimensionality reduction, a multiclass KSVC is trained which can estimate the view of head. Given a testing image x_{ts}, it is first mapped by the kernel and then the low-dimensional features can be obtained by the projection matrix learned from KPCA $y_{ts} = \overline{\mathbf{V}}^T\phi(x_{ts})$ which will be fed into the KSVC to predict the head pose estimation. This method is proved to be outperformed its linear counterpart, i.e., PCA + SVC.

4.2. View representation by Isomap

The derivation of how the Isomap reduces the dimensionality of the original data to a low-dimension has been introduced in the previous section. Now the problem is how to connect the head pose to the features. A pose parameter map **F** is proposed in Ref. [22] to build such connections.

$$\mathbf{\Theta} = \mathbf{FY} \tag{12}$$

where $\mathbf{\Theta} = (\boldsymbol{\theta}_1, \boldsymbol{\theta}_2, \ldots, \boldsymbol{\theta}_M)$ denotes the angles of the head poses from the training data and **Y** is the low-dimensional representation of the data obtained from Isomap. Actually, the matrix of **F** can be seen as a set of linear transformations that map the features to corresponding pose angles. During training time, the head poses $\mathbf{\Theta}$ are given as annotations, and the low-dimensional features **Y** can be learned by manifold learning, then, **F** can be obtained using the singular value decomposition (SVD) of $\mathbf{Y}^T$.

$$\mathbf{F}^T = \mathbf{P_Y}\mathbf{W_Y^{-1}}\mathbf{U_Y^T}\mathbf{\Theta}^T \tag{13}$$

where $\mathbf{P_Y}$, $\mathbf{W_Y}$, and $\mathbf{U_Y}$ are the SVD of $\mathbf{Y}^T$.

Given a testing image x_{ts}, the goal now is to obtain the low-dimensional feature according to the embedding **Y**. The first step is to construct a geodesic distance vector for the testing image to all the training images $\boldsymbol{d}_{ts,M} = (d^2_{ts,1}, d^2_{ts,2}, \ldots, d^2_{ts,M},)^T$. Then, $\boldsymbol{d}_{ts} = diag(\mathbf{Y}^T\mathbf{Y}) - \boldsymbol{d}_{ts,M}$. Next, the low-dimensional representation of the testing image is obtained by

$$y_{ts} = \frac{1}{2}\left((\mathbf{Y}^T\mathbf{Y})^{-1}\mathbf{Y}^T\right)^T \boldsymbol{d}_{ts} \tag{14}$$

Finally, the estimated pose of the testing image is computed from

$$\boldsymbol{\theta}_{ts} = \mathbf{F}y_{ts} \tag{15}$$

The insight of this method focuses on the conversion from testing data to the subspace learned by nonlinear manifold learning. The algorithms of LLE and LE can also be generalized by the proposed idea.

4.3. The biased manifold embedding (BME)

The head pose estimation is subjected to the identity variation. The ideal case is to eliminate such negative effects, which means the face images with close pose angles should maintain nearer and the ones with quite different poses should stay farther in the low-dimensional manifold, even the poses are from the same identity. Based on this statement, the BME is proposed to modify the distance matrix according to the pose angles, which can be extended with almost all the classical algorithms [23].

The modified distance between a pair of data points $\boldsymbol{x}_i$ and $\boldsymbol{x}_j$ is given by:

$$\tilde{d}_{ij} = \begin{cases} \dfrac{\beta * p(i,j)}{\max_{m,n}\left(p(m,n)\right) - p(i,j)} * d_{ij}, & p(i,j) \neq 0 \\ 0, & p(i,j) = 0 \end{cases} \tag{16}$$

where $p(i,j) = |p_i - p_j|$ is simply defined as the absolute difference of the angles of two poses. From the modified distance matrix, one can find that the distance between images with close poses is biased to be proportionally small. The images with the same poses are defined to be zero-distance.

In fact, the BME can be seen as a naÏve version of the supervised manifold learning. The head pose information is used as the supervision to enhance the construction of the graph. For the head pose estimation stage, the generalized regression neural network (GRNN) [24] is applied to learn the nonlinear mapping for the unseen data points, and linear multivariate regression is applied to estimate the head pose angle. This idea can be easily extended to the classical algorithms, e.g., Isomap, LLE, and LE, among which the biased LE achieves the lowest error rate on the data set of FacePix [25].

4.4. Head pose estimation as frontal view search

The two remarkable head poses, i.e., yaw and pitch, cause the problem self-occlusion. Compared with pitch, the yaw makes the problem more serious. An extended manifold learning (EML) method is proposed to specify the head pose estimation only considering the variation in the yaw [15]. This work resorts to the frontal view search instead of directly estimating the head pose, which is more efficient and robust. The idea is based on the observation that the frontal face locates nearly at the vertex in the symmetrical shape of the embedding. However, if the pose distribution of the data is asymmetric, the location of the frontal face in the manifold will shift from the vertex. Therefore, the first trial of the EML method is data enhancement. All the images are horizontally flipped and both the original and flipped images are used for manifold learning. In order to make the method more robust to variations in environment, for example, illumination, the localized edge orientation histogram (LEOH) is presented to represent the original color mappings as more representative features. The idea is inspired by the classical HoG feature [14]. The first step of LEOH is to apply a Canny edge detector on the original images. Then, the whole image is divided into $M \times N$ cells. The gradient orientation

http://dx.doi.org/10.5772/65903

is quantized into N_B bins. Next, histograms of the gradient orientation of the cells locating in a block consisted of $P \times Q$ cells are accumulated and normalized. Finally, the LEOH feature is obtained by the block features concatenation. The proposed ideas can be easily incorporated in various manifold learning methods that improve the performance of the frontal view searching.

4.5. Head pose estimation by supervised manifold learning

A taxonomy of methods, which structures the general framework of manifold learning into several stages, is proposed to incorporate the head pose angles in one or some of the stages to enable the supervised manifold learning [26]. A straightforward solution could be the adaption of the distance and weight matrix according to the angle difference between pairwise face images. The head pose estimation problem is then interrupted as a regression problem, which was usually solved as a classification problem. As a result, continuous head poses can be generalized by the model.

The general framework of manifold learning can be represented as follows: Stage 1, neighborhood searching; Stage 2, graph weighting; Stage 3, low-dimensional manifold computation; and Stage 4, projection from unseen data to the manifold and pose estimation.

In Stage 1, the distance matrix of $\mathbf{D} = \{d_{ij}\}$ can be adapted as follows:

$$\tilde{d}_{ij} = f(|\theta_i - \theta_j|) \cdot d_{ij} \tag{17}$$

where θ_i and θ_j are the angles of two poses, which keep the same denotation as previous sections. The f is some reciprocal increasing positive function, for example, $f(u) = \alpha \cdot u/(\beta - u)$. The introduction of f encourages the distance decreasing of the nearer poses and increasing of farther poses. The farther the poses are, the more penalties the distance will gain.

In Stage 2, the weight matrix of $\mathbf{W} = \{w_{ij}\}$ can be adapted by similar idea of supervision information incorporation.

$$\tilde{w}_{ij} = w_{ij} \cdot g(|\theta_i - \theta_j|) \tag{18}$$

where g is defined as some positive decreasing function, which is similar to the f applied in Stage 1.

In Stage 3, let us take the LLE for an instance. The original objective function of LLE shown in Eq. (6) can be adapted as follows:

$$\min_{\mathbf{Y}} \sum_{i=1}^{M} \left\| \mathbf{y}_i - \sum_{j \in N(i)} w_{ij} \mathbf{y}_j \right\|^2 + \lambda \frac{1}{2} \sum_{i,j} (\mathbf{y}_i - \mathbf{y}_j)^2 \Lambda_{ij} \tag{19}$$

where the $\mathbf{\Lambda} = \{\Lambda_{i,j}\}$ measures the similarity between the angles of pairwise poses. A possible form of Λ is the heat kernel.

$$\Lambda_{ij} = \begin{cases} e^{-\frac{\left\|\theta_i - \theta_j\right\|^2}{2\sigma^2}}, & \text{if the } i^{th} \text{ and } j^{th} \text{data points are neighbours} \\ 0, & \text{otherwise} \end{cases} \tag{20}$$

The adaption of the objective function can preserve the local linearity of the original data and enhance the similarity for neighborhoods, which are facilitated with similar poses. This is implemented by the second term of Eq. (19) that introduces the supervision information. Following the derivation from Eq. (6) to Eq. (7), Eq. (19) can be simplified as:

$$\min_{\mathbf{Y}} \mathbf{YMY}^T + \lambda \mathbf{Y\tilde{L}Y}^T = \min_{\mathbf{Y}} \mathbf{Y}(\mathbf{M} + \lambda\tilde{\mathbf{L}})\mathbf{Y}^T \tag{21}$$

where $\tilde{\mathbf{L}}$ is the Laplacian matrix of $\mathbf{\Lambda}$. For the low-dimensional embedding, eigenvectors decomposition of $\mathbf{M} + \lambda\mathbf{L}$ can be performed. By the supervision information incorporation, the method is much capable of imposing discriminative projection to the learned embedding.

In Stage 4, the GRNN algorithm is applied to produce the mapping from unseen data to the low-dimensional embedding. During testing time, the support vector regression (SVR) with RBF kernel and smoothing cubic splines are taken.

A novel method of supervised manifold learning for head pose estimation [27, 28] is proposed based on the framework from the former method. Similarly, angles of poses are incorporated in all three stages of the general manifold learning structure.

In Stage 1, an improved version of f is proposed as:

$$\tilde{d_{ij}} = f(|\theta_i - \theta_j|)^p \cdot d_{ij}(p > 0) \tag{22}$$

where f is defined as a rectified reciprocal form $f(|\theta_i - \theta_j|) = \alpha \frac{|\theta_i - \theta_j|}{\max_{m,n}\{|\theta_m - \theta_n|\} - |\theta_i - \theta_j| + \varepsilon}$. α is a positive constant and ε is an arbitrary small positive constant that avoids the denominator of f being zero. This adaption for the distance matrix further enhances the effects of the supervision information during the procedure of neighbors search.

In Stage 2, taking LLE (NPE [29]), for an example, the local distance matrix shown in Eq. (4) is modified as

$$\tilde{c_{mn}} = g_{mn} \cdot c_{mn} \tag{23}$$

where $g_{mn} = \frac{|\theta_i - \theta_m||\theta_i - \theta_n|}{(\max_{m,n}\{|\theta_m - \theta_n|\} - |\theta_m - \theta_n| + \varepsilon)^2}$. θ_i is the angle of the reference face image x_i.This operation enhances the supervision during the computation of local correlated matrix.

In Stage 3, a supervised neighborhood-based fisher discriminant analysis (SNFDA) is proposed. The basic idea is to make the neighboring data points as close as possible and the nonneighboring data points as far as possible in the low-dimensional embedding. The SNFDA can be seen as a postprocessing procedure in this stage. Based on the low-dimensional represented data $\mathbf{Y}$ obtained from the original LLE or the modified LLE in Stages 1 and 2, the within- and between-neighborhood scatter matrices are defined as:

$$\mathbf{S}_w = \frac{K}{2}\sum_{i,j=1}^{M} A_{ij}^{w}(\boldsymbol{y}_i - \boldsymbol{y}_j)(\boldsymbol{y}_i - \boldsymbol{y}_j)^T \tag{24}$$

$$\mathbf{S}_B = \frac{K}{2}\sum_{i,j=1}^{M} A_{ij}^{B}(\boldsymbol{y}_i - \boldsymbol{y}_j)(\boldsymbol{y}_i - \boldsymbol{y}_j)^T \tag{25}$$

where

$$A_{ij}^{w} = \begin{cases} \frac{A_{ij}}{K}, & \boldsymbol{y}_j \in N(i) \\ 0, & \text{otherwise} \end{cases} \tag{26}$$

$$A_{ij}^{w} = \begin{cases} A_{ij}\left(\frac{1}{M} - \frac{1}{K}\right), & \boldsymbol{y}_j \in N(i) \\ \frac{A_{ij}}{n}, & \text{otherwise} \end{cases} \tag{27}$$

A_{ij} is the affinity between $\boldsymbol{y}_i$ and $\boldsymbol{y}_j$, which is defined as the form of heat function:

$$A_{ij} = e^{-\frac{\|\boldsymbol{y}_i - \boldsymbol{y}_j\|^2}{2\sigma^2}} \tag{28}$$

Details about the inference of the scatter matrices can be found in the original reference. The transformed matrix $\mathbf{T}_{\mathrm{SNFDA}}$ of SNFDA is computed from the generalized eigenvector decomposition problem

$$\mathbf{S}_B \boldsymbol{e} = \lambda \mathbf{S}_w \boldsymbol{e} \tag{29}$$

The top d eigenvectors span to the $\mathbf{T}_{\mathrm{SNFDA}}$ and the transformed feature is obtained by $z = \mathbf{T}_{\mathrm{SNFDA}}\mathbf{y}$. This supervised learning manner successfully introduces the supervision information in a framework to provide a "good" projection from the original data to the low-dimensional. Due to the supervised learning, when the projection is applied on original data, more discriminative features can be obtained for head pose estimation.

In Stage 4, during testing time, the GRNN is applied to map the unseen data point to the low-dimensional embedding and the relevance vector machine (RVM) [30] is adopted to accomplish the pose estimation. Experimental results obtained by the proposed method performing on the database of FacePix [25] and MIT-CBCL [31] show big improvements compared with other state-of-the-art algorithms [23, 26] in Stage 3 and Stages 1 + 2 + 3. This means that this method is more robust for identity and illumination variations.

5. Summary

In this chapter, the head pose estimation, one of the most challenging tasks in the area of computer vision, is introduced, and the main types of methods are demonstrated and compared. Particularly, the manifold learning-based methods are attracted more attention. In reality, data distribution is usually nonlinear in high-dimensional represented space, e.g., the head pose images. Some potential structures are lying on nonlinear but smooth manifolds which are embedded in the original space. The manifold learning algorithms are able to discover and visualize such embedding. Almost all the algorithms are formalized based on the assumption of the local linearity of the nonlinear data. Those algorithms highly benefit the application of head pose estimation, because the face orientations (yaw and pitch) are found to be distributed along some specific manifolds. Promising performance is achieved by the classical manifold learning methods, which, however, are highly improved by the supervised manifold learning. It proves that the supervised information represented as angles of head poses is helpful in head pose estimation. However, there are still hurdles to take. Most of the methods are tested in different settings, e.g., different database is used in different method. A common framework could help to offer fair justifications. Other feature instead of the simple color space can be considered to better represent the face images. Some useful tools are available online to help better understand the work [16, 32, 33].

Author details

Chao Wang[1,†*], Yuanhao Guo[2,†] and Xubo Song[3]

*Address all correspondence to: chaocharleswang@gmail.com

1 Online Search Team, The Home Depot, Atlanta, GA, USA

2 Imaging & BioInformatics, Leiden Institute of Advanced Computer Science, Leiden University, Leiden, The Netherlands

3 Department of Biomedical Engineering, Oregon Health & Science University, Portland, OR, USA

† These authors contributed equally to this work

References

[1] Murphy-Chutorian, Erik and Trivedi, Mohan Manubhai. Head pose estimation in computer vision: A survey. IEEE Transactions on Pattern Analysis and Machine Intelligence. 2009;**31**(4):607–626.

[2] Gourier, Nicolas, Hall, Daniela and Crowley, James L. Estimating face orientation from robust detection of salient facial structures. In: 2004 ICPR Workshop on Visual Observation of Deictic Gestures. Cambridge, UK: IEEE; 2004.

[3] Taigman, Yaniv, Yang, Ming, Ranzato, Marc'Aurelio and Wolf, Lior. Deepface: Closing the gap to human-level performance in face verification. In: 2014 IEEE Conference on Computer Vision and Pattern Recognition. Columbus, OH, USA: IEEE; 2014. pp. 1701–1708.

[4] Beymer, David James. Face recognition under varying pose. In: 1994 IEEE Conference on Computer Vision and Pattern Recognition. Seattle, WA, USA: IEEE; 1994. pp. 756–761.

[5] Niyogi, Sourabh and Freeman, William T. Example-based head tracking. In: 2nd IEEE International Conference on Automatic Face and Gesture Recognition. Killington, VT, USA: IEEE; 1996. pp. 374–378.

[6] Viola, Paul and Jones, Michael. Rapid object detection using a boosted cascade of simple features. In: 2001 IEEE Conference on Computer Vision and Pattern Recognition. Kauai, HI, USA: IEEE; 2001. pp. I-511–I-518.

[7] Felzenszwalb, Pedro F and Huttenlocher, Daniel P. Pictorial structures for object recognition. International Journal of Computer Vision. 2005;**61**(1):55–79.

[8] Cootes, Timothy F, Edwards, Gareth J and Taylor, Christopher J. Active appearance models. In: 1998 European Conference on Computer Vision. Freiburg, Germany: Springer; 1998. pp. 484–498.

[9] Matthews, Iain and Baker, Simon. Active appearance models revisited. International Journal of Computer Vision. 2004;**60**(2):135–164.

[10] Tenenbaum, Joshua B, De Silva, Vin and Langford, John C. A global geometric framework for nonlinear dimensionality reduction. Science. 2000;**290**(5500):2319–2323.

[11] Roweis, Sam T and Saul, Lawrence K. Nonlinear dimensionality reduction by locally linear embedding. Science. 2000;**290**(5500):2323–2326.

[12] Belkin, Mikhail and Niyogi, Partha. Laplacian eigenmaps for dimensionality reduction and data representation. Neural Computation. 2003;**15**(6):1373–1396.

[13] Maaten, Laurens van der and Hinton, Geoffrey. Visualizing data using t-SNE. Journal of Machine Learning Research. 2008;**9**:2579–2605.

[14] Dalal, Navneet and Triggs, Bill. Histograms of oriented gradients for human detection. In: IEEE Conference on Computer Vision and Pattern Recognition; San Diego, CA, USA: IEEE; 2005. pp. 886–893.

[15] Wang, Chao and Song, Xubo. Robust frontal view search using extended manifold learning. Journal of Visual Communication and Image Representation. 2013;**24**(7):1147–1154.

[16] Available from: http://isomap.stanford.edu/

[17] Balasubramanian, Mukund and Schwartz, Eric L. The isomap algorithm and topological stability. Science. 2002;295(5552):7–7.

[18] Cox, Trevor F and Cox, Michael AA. Multidimensional Scaling. USA: CRC Press; 2000.

[19] Weinberger, Kilian Q and Saul, Lawrence K. Unsupervised learning of image manifolds by semidefinite programming. International Journal of Computer Vision. 2006;**70**(1):77–90.

[20] Sherrah, Jamie, Gong, Shaogang and Ong, Eng-Jon. Face distributions in similarity space under varying head pose. Image and Vision Computing. 2001;**19**(12):807–819.

[21] Li, Stan Z, Fu, Qingdong, Gu, Lie, Scholkopf, B, Cheng, Yimin and Zhang, Hongjiag. Kernel machine based learning for multi-view face detection and pose estimation. In: 2001 IEEE International Conference on Computer Vision. Vancouver, Canada: IEEE; 2001.

[22] Raytchev, Bisser, Yoda, Ikushi and Sakaue, Katsuhiko. Head pose estimation by nonlinear manifold learning. In: The 17th International Conference on Pattern Recognition; Cambridge, UK: IEEE; 2004. pp. 462–466.

[23] Balasubramanian, Vineeth Nallure, Ye, Jieping and Panchanathan, Sethuraman. Biased manifold embedding: A framework for person-independent head pose estimation. In: 2007 IEEE Conference on Computer Vision and Pattern Recognition. Minneapolis, MN, USA: IEEE; 2007. pp. 1–7.

[24] Specht, Donald F. A general regression neural network. IEEE Transactions on Neural Network. 1991;**2**(6):568–576.

[25] Little, Danny, Krishna, Sreekar, John Jr, A and Panchanathan, Sethuraman. A methodology for evaluating robustness of face recognition algorithms with respect to variations in pose angle and illumination angle. In: ICASSP (2); Citeseer; 2005. pp. 89–92.

[26] BenAbdelkader, Chiraz. Robust head pose estimation using supervised manifold learning. In: European Conference on Computer Vision; Crete, Greece: Springer; 2010. p. 518–531.

[27] Wang, Chao and Song, Xubo. Robust head pose estimation via supervised manifold learning. Neural Networks. 2014;**53**:15–25.

[28] Wang, Chao and Song, Xubo. Robust head pose estimation using supervised manifold projection. In: 2012 19th IEEE International Conference on Image Processing. Orlando, FL, USA: IEEE; 2012. pp. 161–164.

[29] He, Xiaofei, Cai, Deng, Yan, Shuicheng and Zhang, Hong-Jiang. Neighbourhood preserving embedding. In: 2005 IEEE International Conference on Computer Vision. Beijing, China: IEEE; 2005. pp. 1208–1213.

[30] Tipping, Michael E. Sparse Bayesian learning and the relevance vector machine. Journal of Machine Learning Research. 2001;**12**:11–244.

[31] Huang, Jennifer, Heisele, Bernd and Blanz, Volker. Component-based face recognition with 3D morphable models. In: International Conference on Audio-and Video-Based Biometric Person Authentication. Guildford, UK: Springer; 2003. pp. 27–34.

[32] Available from: https://www.cs.nyu.edu/~roweis/lle/code.html

[33] Available from: https://lvdmaaten.github.io/drtoolbox/

Chapter 7

A Fusion Scheme of Local Manifold Learning Methods

Xianglei Xing, Kejun Wang and Weixing Feng

Additional information is available at the end of the chapter

http://dx.doi.org/10.5772/66303

Abstract

Spectral analysis-based dimensionality reduction algorithms, especially the local manifold learning methods, have become popular recently because their optimizations do not involve local minima and scale well to large, high-dimensional data sets. Despite their attractive properties, these algorithms are developed based on different geometric intuitions, and only partial information from the true geometric structure of the underlying manifold is learned by each method. In order to discover the underlying manifold structure more faithfully, we introduce a novel method to fuse the geometric information learned from different local manifold learning algorithms in this chapter. First, we employ local tangent coordinates to compute the local objects from different local algorithms. Then, we utilize the truncation function from differential manifold to connect the local objects with a global functional and finally develop an alternating optimization-based algorithm to discover the low-dimensional embedding. Experiments on synthetic as well as real data sets demonstrate the effectiveness of our proposed method.

Keywords: dimensionality reduction, manifold learning

1. Introduction

Nonlinear dimensionality reduction (NLDR) plays an important role in the modern data analysis system, since many objects in our world can only be electronically represented with high-dimensional data such as images, videos, speech signals, and text documents. We usually need to analyze a large amount of data and process them, and however, it is very complicated or even infeasible to process these high-dimensional data directly, due to their high computational complexity on both time and space. Over the past decade, numerous manifold learning methods have been proposed for nonlinear dimensionality reduction. From methodology, these methods can be divided into two categories: global algorithms and local algorithms. Representative global algorithms contain isometric mapping [1], maximum variance unfolding

[2], and local coordinates alignment with global preservation [3]. Local methods mainly include Laplacian eigenmaps (LEM) [4], locally linear embedding (LLE) [5], Hessian eigenmaps (HLLE) [6], local tangent space alignment (LTSA) [7], local linear transformation embedding [8], stable local approaches [9], and maximal linear embedding [10].

Different local approaches try to learn different geometric information of the underlying manifold, since they are developed based on the knowledge and experience of experts for their own purposes [11]. Therefore, only partial information from the true underlying manifold is learned by each existing local manifold learning method. Thus, to better discover the underlying manifold structure, it is more informative and essential to provide a common framework for synthesizing the geometric information extracted from different local methods. In this chapter, we propose an interesting method to unify the local manifold learning algorithms (e. g., LEM, LLE, HLLE, and LTSA). Inspired by HLLE which employs local tangent coordinates to compute the local Hessian, we propose to utilize local tangent coordinates to estimate the local objects defined in different local methods. Then, we employ the truncation function from differential manifold to connect the local objects with a global functional. Finally, we develop an alternating optimization-based algorithm to discover the global coordinate system of lower dimensionality.

2. Local tangent coordinates system

A manifold is a topological space that locally resembles Euclidean space near every point. For example, around each point, there is a neighborhood that is topologically the same as the open unit ball in $\mathbb{R}^D$. The simplest manifold is a linear manifold, usually called a hyperplane. There exists a tangent space at each point of a nonlinear manifold. The tangent space is a linear manifold which locally approximates the manifold. Suppose there are N points $\{x_1,\ldots,x_N\}$ in $\mathbb{R}^D$ residing on a smooth manifold $\mathcal{M}\subset\mathbb{R}^D$, which is the image of a coordinate space $\mathcal{Y}\subset\mathbb{R}^d$ under a smooth mapping $\psi:\mathcal{Y}\to\mathbb{R}^D$, where $d\ll D$. The mapping ψ is assumed as a locally isometric embedding. The aim of a NLDR algorithm is to acquire the corresponding low-dimensional representation $y_i\in\mathcal{Y}$ of each $x_i\in\mathcal{M}$ and preserve certain intrinsic structures of data at the same time. Suppose $\mathcal{M}$ is smooth such that the tangent space $T_x(\mathcal{M})$ is well defined at every point $x\in\mathcal{M}$. We can regard the local tangent space as a d-dimensional affine subspace of $\mathbb{R}^D$ which is tangent to $\mathcal{M}$ at x. Thus, the tangent space has the natural inner product induced by the embedding $\mathcal{M}\subset\mathbb{R}^D$. Within some neighborhood of x, each point $x\in\mathcal{M}$ has a sole closest point in $T_x(\mathcal{M})$, and therefore, an orthonormal coordinate system from the corresponding local coordinates on $\mathcal{M}$ can be associated with the tangent space.

A manifold can be represented by its coordinates. While the current research of differential geometry focuses on the characterization of the global properties of manifolds, NLDR algorithms, which try to find the coordinate representations of data, only need the local properties of manifolds. In this chapter, we use local coordinates associated with the tangent space to estimate the local objects over the manifold. To acquire the local tangent coordinates, we first perform Principal Component Analysis (PCA) [12] on the points in $\mathcal{N}(x_i)=\{x_i,x_{i_1},\ldots,x_{i_k}\}$ that is the local patch built by the point x_i and its k nearest neighborhoods, and get d leading PCA eigenvectors $V^i = \{v_1^i, v_2^i, \ldots, v_d^i\}$ which correspond to an orthogonal basis of $T_{x_i}(\mathcal{M})$ (the orthogonal basis can be seen as a d-dimensional affine subspace of $\mathbb{R}^D$ which is tangent to $\mathcal{M}$ at x_i). For high-dimensional data, we employ the trick presented by Turk and Pentland for EigenFaces [13]. Then, we obtain the local tangent coordinates $\mathcal{U}^i = \{0, u_1^i, \ldots, u_k^i\}$ of the neighborhood $\mathcal{N}(x_i)$ by projecting the local neighborhoods to this tangent subspace:

$$u_j^i = (V^i)^T (x_{i_j} - x_i) \tag{1}$$

An illustration of the local tangent space at x_i and the corresponding tangent coordinates system (i.e., the point x_{i_j}'s local tangent coordinate is u_j^i) is shown in **Figure 1**.

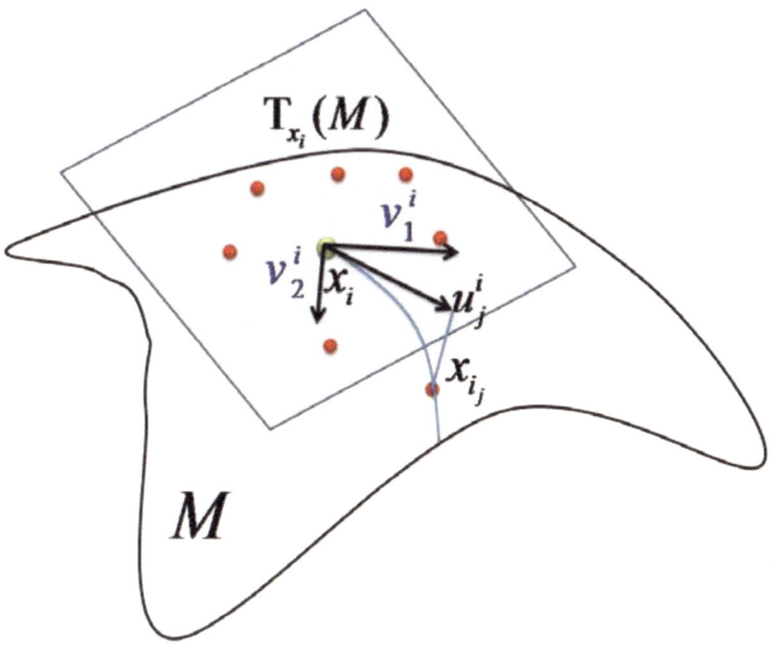

Figure 1. Local tangent space and tangent coordinates system.

3. Reformulations of LEM, LLE, HLLE and LTSA using local tangent coordinates

3.1. Reformulation of Laplacian eigenmaps

The method LEM was introduced by Belkin and Niyogi [4]. We can summarize the geometrical motivation of LEM as follows. Assume that we are searching for a smooth one-dimensional embedding $f : \mathcal{M} \rightarrow \mathbb{R}$ from the manifold to the real line so that data points near each together on the manifold are also mapped close together on the line. Think about two adjacent points, $x,z \in \mathcal{M}$, which are mapped to $f(x)$ and $f(z)$, respectively, we can obtain that

$$|f(z) - f(x)| \leq \|\nabla_{\mathcal{M}} f(x)\| \|z - x\| + O(\|z - x\|^2) \tag{2}$$

where $\nabla_{\mathcal{M}} f$ is the gradient vector field along the manifold. Thus, to the first order, $\|\nabla_{\mathcal{M}} f\|$ provides us with an estimate of how far apart f maps nearby points. When we look for a map that best preserves locality on average, a natural choice to find f is to minimize [4]:

$$\Phi_{lap}(f) = \int_{\mathcal{M}} \|\nabla_{\mathcal{M}} f\|^2 = \int_{\mathcal{M}} \Delta_{\mathcal{M}}(f) f \tag{3}$$

where the integral is taken with respect to the standard measure over the manifold. Thus, the function f that minimizes $\Phi_{lap}(f)$ has to be an eigenfunction of the Laplace-Beltrami operator $\Delta_{\mathcal{M}}$, which is a key geometric object associated with a Riemannian manifold [14].

Suppose that the tangent coordinate of $x \in \mathcal{N}(x)$ is given by u. Then, the rule $g(u) = f(x) = f \circ \psi(u)$ defines a function $g: U \to \mathbb{R}$, where U is the neighborhood of $u \in \mathbb{R}^d$. With the help of local tangent coordinates, we can reduce the computation of the gradient vector $\nabla_{\mathcal{M}} f(x)$ on the manifold to the computation of the ordinary gradient vector on the Euclidean space:

$$\nabla_{tan} f(x) = \nabla g(u) = \left(\frac{\partial g(u)}{\partial u^1}, \ldots, \frac{\partial g(u)}{\partial u^d} \right)^T \tag{4}$$

where $u = (u^1, \ldots, u^d) \in \mathbb{R}^d$, and we keep up *tan* in the notation to make clear that it counts on the coordinate system in $T_x(\mathcal{M})$. For different local coordinate systems, although the tangent gradient vector will be different, the norm $\|\nabla_{tan} f(x)\|$ is inimitably defined such that equation (3) can be approximated by estimating the following functional:

$$\tilde{\Phi}_{lap}(f) = \int_{\mathcal{M}} \|\nabla_{tan} f(x)\|^2 dx \tag{5}$$

where dx stands for the probability measure on $\mathcal{M}$.

In order to compute the local object $\|\nabla_{tan} f(x)\|^2$, we first use the first-order Taylor series expansion to approximate the smooth functions $\{f(x_{i_j})\}_{j=1}^k$, $f: \mathcal{M} \to \mathbb{R}$, and together with Eq. (4), we have:

$$\begin{aligned} f(x_{i_j}) &= f(x_i) + (\nabla_{tan} f(x_i))^T (x_{i_j} - x_i) + O(\|x_{i_j} - x_i\|^2) \\ &= g(u_j^i) = g(0) + (\nabla_{tan} f(x_i))^T u_j^i + O(\|u_j^i\|^2) \end{aligned} \tag{6}$$

Over $\mathcal{U}^i$, we develop the operator $\alpha^i = [g(0), \nabla g(0)] = [g(0), \nabla_{tan} f(x_i)]$ that approximates the function $g(u_j^i)$ by its projection on the basis $U_j^i = \{1, u_{j_1}^i, \ldots, u_{j_d}^i\}$:

$$f(x_{i_j}) = g(u_j^i) = (\alpha^i)^T U_j^i \tag{7}$$

The least-squares estimation of the operator α^i can be computed by:

$$\underset{\alpha^i}{\text{argmin}} \sum_{j=1}^{k} (f(x_{i_j}) - (\alpha^i)^T U_j^i)^2 \tag{8}$$

It is easy to show that the least-squares solution of the above object function is $\alpha^i = (U^i)^{\dagger} f^i$, where $f^i = [f(x_{i_1}), \ldots, f(x_{i_k})] \in \mathbb{R}^k$, $U^i = [U_1^i; U_2^i; \ldots; U_k^i] \in \mathbb{R}^{k \times (1+d)}$, and $(U^i)^{\dagger}$ denotes the pseudo-inverse of U^i. If we define a local gradient operator $G^i \in \mathbb{R}^{d \times k}$ which is constructed by the last d rows of $(U^i)^{\dagger}$, we have $\nabla_{tan} f(x_i) = G^i f^i$. Furthermore, the local object $\|\nabla_{tan} f(x_i)\|^2$ can be computed as:

$$\|\nabla_{tan} f(x_i)\|^2 = \nabla_{tan} f(x_i)^T \nabla_{tan} f(x_i) = (f^i)^T (G^i)^T G^i f^i \tag{9}$$

An unresolved problem in our reformulation is how to connect the local object $\|\nabla_{tan} f(x)\|^2$ with the global functional $\tilde{\Phi}_{lap}(f)$ in (5) and its discrete approximation. In Section4, we will discuss this issue in detail.

3.2. Reformulation of locally linear embedding

The LLE method was introduced by Roweis and Saul [5]. It is based on simple geometric intuitions, which can be depicted as follows. Globally, the data points are sampled from a nonlinear manifold, while each data point and its neighbors are residing on or close to a linear patch of the manifold locally. Thus, it is possible to describe the local geometric properties of the neighborhood of each data point in the high-dimensional space by linear coefficients which reconstruct the data point from its neighbors under suitable conditions. The method of LLE computes the low-dimensional embedding which is optimized to preserve the local configurations of the data. In each locally linear patch, the reconstruction error in the original LLE can be written as:

$$\hat{\varepsilon}^i = \|x_i - \sum_{j=1}^{k} w_{i_j} x_{i_j}\|^2 \tag{10}$$

where $\{w_{i_j}\}_{j=1}^k$ are the reconstruction weights which encode the geometric information of the high-dimensional inputs and are constrained to satisfy $\sum_j w_{i_j} = 1$.

Since the geometric structure of the local patch can be approximated by its projection on the tangent space $T_{x_i}(\mathcal{M})$, we utilize the local tangent coordinates to estimate the local objects over the manifold in our reformulation framework. We can write the reconstruction error of each local tangent coordinate as:

$$\varepsilon^i = \|u_i - \sum_{j-1}^{k} w_{i_j} u_j^i\|^2 = \|\sum_j w_{i_j}(u_i - u_j^i)\|^2 = \sum_{jk} w_{i_j} w_{i_k} G_{jk}^i \tag{11}$$

where we have employed the fact that the weights sum to one, and G^i is the local Gram matrix,

$$G_{jk}^i = \langle (u_i - u_j^i), (u_i - u_k^i) \rangle \tag{12}$$

The optimal weights can be obtained analytically by minimizing the above reconstruction error. We solve the linear system of equations

$$\sum_k G_{jk}^i w_{i_k} = 1 \tag{13}$$

and then normalize the solution by $\sum_k w_{i_k} = 1$. Consider the problem of mapping the data points from the manifold to a line such that each data point on the line can be represented as a linear combination of its neighbors. Let $f(x_{i_1}),\ldots,f(x_{i_k})$ denote the mappings of $u_1^i,\ldots,u_k^i$, respectively. Motivated by the spirit of LLE, the neighborhood of $f(x_i)$ should share the same geometric information as the neighborhood of u_i, so we can define the following local object:

$$|\sigma_f(x_i)|^2 = |f(x_i) - \sum_{j=1}^{k} w_{i_j} f(x_{i_j})|^2 = (f^i)^T (W^i)^T W^i f^i \tag{14}$$

where $W^i = [1, -w_i] \in \mathbb{R}^{1\times(k+1)}, f^i = [f(x_i), f(x_{i_1}), \ldots, f(x_{i_k})]$. The optimal mapping f can be obtained by minimizing the following global functional:

$$\mathcal{E}(f) = \int_{\mathcal{M}} |\sigma_f(x)|^2 dx \tag{15}$$

where dx stands for the probability measure on the manifold.

3.3. Reformulation of Hessian eigenmaps

The HLLE method was introduced by Donoho and Grimes [6]. In contrast to LLE that obtains linear embedding by minimizing the l_2 error in Eq. (10), the HLLE achieves linear embedding by minimizing the Hessian functional on the manifold where the data points reside. HLLE supposes that we can obtain the low-dimensional coordinates from the $(d+1)$-dimensional null-space of the functional $\mathcal{H}(f)$ which presents the average curviness of f upon the manifold, if the manifold is locally isometric to an open connected subset of $\mathbb{R}^d$. We can measure the functional $\mathcal{H}(f)$ by averaging the Frobenius-norm of the Hessians on the manifold $\mathcal{M}$ as [6]:

$$\mathcal{H}(f) = \int_{\mathcal{M}} \|H_f^{tan}(x)\|_F^2 dx \tag{16}$$

where H_f^{tan} stands for the Hessian of f in tangent coordinates. In order to estimate the local Hessian matrix, we first perform a second-order Taylor expansion at a fixed x_i on the smooth functions: $\{f(x_{i_j})\}_{j=1}^k, f : \mathcal{M} \to \mathbb{R}$ that is C^2 near x_i:

$$
\begin{aligned}
f(x_{i_j}) &\approx f(x_i) + (\nabla f)^T(x_{i_j} - x_i) + \frac{1}{2}(x_{i_j} - x_i)^T H_f^i (x_{i_j} - x_i) \\
&= g(u_j^i) = g(0) + (\nabla g)^T u_j^i + \frac{1}{2} u_j^{iT} H_f^i u_j^i + O(\|u_j^i\|^3)
\end{aligned}
\tag{17}
$$

Here, $\nabla f = \nabla g$ is the gradient defined in (4), and H_f^i is the local Hessian matrix defined as:

$$
(H_f^i)_{p,q}(x) = \frac{\partial}{\partial u_p}\frac{\partial}{\partial u_q} g(u) \tag{18}
$$

where $g : U \to \mathbb{R}$ uses the local tangent coordinates and satisfies the rule $g(u) = f(x) = f \circ \psi(u)$. In the second identity of Eq. (17), we have exploited the fact that $u_i^i = \langle V^i, x_i - x_i \rangle = 0$ [recall the computation of local tangent coordinates in Eq. (1)].

Over $\mathcal{U}^i$, we develop the operator β^i that approximates the function $g(u_j^i)$ by its projection on the basis $U_j^i = \{1, u_{j_1}^i, \ldots, u_{j_d}^i, (u_{j_1}^i)^2, \ldots, (u_{j_d}^i)^2, \ldots, u_{j_1}^i \times u_{j_2}^i, \ldots, u_{j_{d-1}}^i \times u_{j_d}^i\}$, and we have:

$$
f(x_{i_j}) = g(u_j^i) = (\beta^i)^T U_j^i \tag{19}
$$

Let $\beta^i = [g(0), \nabla g, h^i] \in \mathbb{R}^{1+d+d(d+1)/2}$, then $h^i \in \mathbb{R}^{d(d+1)/2}$ is the vector form of local Hessian matrix H_f^i over neighborhood $N(x_i)$. The least-squares estimation of the operator β^i can be obtained by:

$$
\underset{\beta^i}{\operatorname{argmin}} \sum_{j=1}^{k} (f(x_{i_j}) - (\beta^i)^T U_j^i)^2 \tag{20}
$$

The least-squares solution is $\beta^i = (U^i)^\dagger f^i$, where $f^i = [f(x_1), \ldots f(x_k)] \in \mathbb{R}^k$, $U^i = [U_1^i; U_2^i; \ldots; U_k^i] \in \mathbb{R}^{k \times (1+d+d(d+1)/2)}$, and $(U^i)^\dagger$ signifies the pseudo-inverse of U^i. Notice that h^i is the vector form of local Hessian matrix H_f^i, while the last $d(d+1)/2$ components of β^i correspond to h^i. Meanwhile, we can construct the local Hessian operator $H^i \in \mathbb{R}^{(d(d+1)/2) \times k}$ by the last $d(d+1)/2$ rows of $(U^i)^\dagger$, and therefore, we can obtain $h^i = H^i f^i$. Thus, the local object $\|H_f^{tan}(x_i)\|_F^2$ can be estimated with:

$$
\|H_f^{tan}(x_i)\|_F^2 = (h^i)^T (h^i) = (f^i)^T (H^i)^T (H^i)(f^i) \tag{21}
$$

3.4. Reformulation of local tangent space alignment

The method LTSA was introduced by Zhang and Zha [7]. LTSA is based on similar geometric intuitions as LLE. The neighborhoods of each data point remain nearby and similarly colocated in the low-dimensional space, if the data set is sampled from a smooth manifold. LLE constructs low-dimensional data so that the local linear relations of the original data are preserved, while LTSA constructs a locally linear patch to approximate the tangent space at the point. The coordinates provided by the tangent space give a low-dimensional representation of the patch. From Eq. (6), we can obtain:

$$f(x_{i_j}) = f(x_i) + (\nabla_{tan} f(x_i))^T u_j^i + O(\|u_j^i\|^2) \tag{22}$$

From the above equation, we can discover that there are some relations between the global coordinate $f(x_{i_j})$ in the low-dimensional feature space and the local coordinate u_j^i which represents the local geometry. The LTSA algorithm requires the global coordinates $f(x_{i_j})$ that should respect the local geometry determined by the u_j^i:

$$f(x_{i_j}) \approx \bar{f}(x_i) + L_i u_j^i, \tag{23}$$

where $\bar{f}(x_i)$ is the mean of $f(x_{i_j})$, $j = 1,\ldots,k$. Inspired by LTSA, the affine transformation L_i should align the local coordinate with the global coordinate, and we can define the following local object:

$$|\kappa_f(x_i)|^2 = |(f^i)^T - \frac{1}{k}(f^i)^T ee^T - L_i U^i|^2, \tag{24}$$

where $f^i = [f(x_{i_1}),\ldots,f(x_{i_k})]^T$, $U^i = [u_1^i; u_2^i; \ldots; u_k^i]$, and e is a k-dimensional column vector of all ones. Naturally, we should seek to find the optimal mapping f and a local affine transformation L_i to minimize the following global functional:

$$\mathcal{K}(f) = \int_{\mathcal{M}} |\kappa_f(x)|^2 dx \tag{25}$$

Obviously, the optimal affine transformation L_i that minimizes the local reconstruction error for a fixed f^i is given by:

$$L_i = (f^i)^T \left(I - \frac{1}{k} ee^T\right) (U^i)^\dagger \tag{26}$$

and therefore,

$$|\kappa_f(x_i)|^2 = |(f^i)^T \left(I - \frac{1}{k} ee^T\right) (I - (U^i)^\dagger U^i)|^2, \tag{27}$$

Let $W^i = (I - (U^i)^\dagger U^i)^T (I - \frac{1}{k} ee^T)^T$, the local object $\kappa_f(x_i)$ can be estimated as:

$$|\kappa_f(x_i)|^2 = |(f^i)^T \left(I - \frac{1}{k} ee^T\right) (I - (U^i)^\dagger U^i)|^2 = (f^i)^T (W^i)^T (W^i)(f^i) \tag{28}$$

4. Fusion of local manifold learning methods

So far we have discussed four basic local objects: $\|\nabla_{tan} f(x)\|^2$, $|\sigma_f(x)|^2$, $\|H_f^{tan}(x_i)\|_F^2$, and $|\kappa_f(x_i)|^2$. From different perspectives, they depict the geometric information of the manifold. We look forward to collect these geometric information together to better reflect the geometric structure of the underlying manifold. Notice that we can estimate these local objects under the local tangent coordinate system according to Eqs. (9), (14), (21), and (28), respectively. Taking stock of the structure of these equations, it is not hard to discover that we can fuse these local objects together under our proposed framework. Assume that there are M different local manifold learning algorithms, we can define the fused local object as follows:

$$LO_f(x) = \sum_{j=1}^{M} c_j LO_j(x) \tag{29}$$

where $\{c_j\}_{j=1}^M$ are the nonnegative balance parameters, $\{LO_j(x)\}_{j=1}^M$ are the local objects, such as $\|\nabla_{tan} f(x)\|^2$, $|\sigma_f(x)|^2$, $\|H_f^{tan}(x_i)\|_F^2$, and $|\kappa_f(x_i)|^2$, from different algorithms. It is worth to note that the other local manifold learning algorithms can also be reformulated to incorporate into our unified framework.

We employ the truncation function from differential manifold to connect the local objects with their corresponding global functional such that we can obtain a consistent alignment of the local objects to discover a single global coordinate system of lower dimensionality. The truncation function is a crucial tool in differential geometry to build relationships between global and local properties of the manifold. Assume that U and V are two nonempty subsets of a smooth manifold $\mathcal{M}$, where $\overline{V}$ is compact and $\overline{V} \in U$ ($\overline{V}$ is the closure of V). Accordingly, the truncation function [15] can be defined as a smooth function $s : \mathcal{M} \to \mathbb{R}$ such that:

$$s(p) = \begin{cases} 1, & p \in V \\ 0, & p \notin U. \end{cases} \tag{30}$$

The truncation function s can be discretely approximated by the 0-1 selection matrix $S^i \in \mathbb{R}^{N\times k}$. An entry of S^i is defined as:

$$(S^i)_{pq} = \begin{cases} 1, & p = N_i\{q\} \\ 0, & p \neq N_i\{q\}. \end{cases} \tag{31}$$

where $N_i = \{i_1,\dots,i_k\}$ denotes the set of indices for the k-nearest neighborhoods of data point x_i. Let $f = [f(x_1),\dots,f(x_N)] \in \mathbb{R}^N$ be a function defined on the whole data set sampled from the global manifold. Thus, the local mapping $f^i = [f(x_1^i),\dots,f(x_k^i)] \in \mathbb{R}^k$ can be expressible by $f^i = (S^i)^T f$. With the help of the selection matrix, we can discretely approximate the global functional $\mathcal{G}(f)$ as follows:

$$\begin{aligned} \mathcal{G}(f) &= \int_{\mathcal{M}} LO_f(x)\,dx = \frac{1}{N}\sum_{i=1}^{N} LO_f(x_i) \\ &= \frac{1}{N}\sum_{i=1}^{N} (f^i)^T \left(\sum_{j=1}^{M} c_j L_j^i\right) f^i = f^T\left(\sum_{j=1}^{M} c_j P^j\right) f \end{aligned} \tag{32}$$

where $\{L_j^i\}_{j=1}^M$ are the local matrices such as $(G^i)^T G^i$, $(W^i)^T W^i$, $(H^i)^T H^i$, and $(W^i)^T W^i$ which are defined in Eqs. (9), (14), (21), and (28). $P^j = \frac{1}{N}\sum_{i=1}^{N} S^i L_j^i (S^i)^T$ is the alignment matrix of the j-th local manifold learning method. The global embedding coordinates $Y = [y_1, y_2, \ldots, y_N] \in \mathbb{R}^{d\times N}$ can be obtained by minimizing the functional $\mathcal{G}(f)$. Let $y = f = [f(x_1), \ldots, f(x_N)]$ be a row vector of Y. It is not hard to show that the global embedding coordinates and the nonnegative weights $c = [c_1, \ldots, c_M]$ can be obtained by minimizing the following objective function:

$$\underset{Y,c}{\operatorname{argmin}} \sum_{j=1}^{M} c_j^r Tr(YP^jY^T)\ s.t. YY^T = I;\ \sum_{j=1}^{M} c_j = 1,\ c_j \geq 0. \tag{33}$$

where the power parameter $r > 1$ is set to avoid the phenomenon that the solution to c is $c_j = 1$ corresponding to the minimum $Tr(YP^jY^T)$ over different local methods and $c_k = 0 (k \neq j)$ otherwise, since our aim is to utilize the complementary geometric information from different manifold learning methods.

We propose to solve the objective function [Eq. (33)] by employing the alternating optimization [16] method, which iteratively updates Y and c in an alternating fashion. First, we fix c to update Y. The optimization problem in Eq. (33) is equivalent to:

$$\underset{Y}{\operatorname{argmin}}\, Tr(YPY^T)\ s.t.\ YY^T = I \tag{34}$$

where $P = \sum_{j=1}^{M} c_j^r P^j$. When c is fixed, we can solve the optimization problem [Eq. (34)] and obtain the global optimal solution Y as the second to $(d+1)$ st smallest eigenvectors of the matrix P. Second, we fix Y to update c. While Y is fixed, we can minimize the objective function [Eq. (33)] analytically through utilizing a Lagrange multiplier to enforce the constraint that $\sum_{j=1}^{M} c_j = 1$. And the global optimal c can be obtained as:

$$c_j = \frac{(1/Tr(YP^jY^T))^{1/(r-1)}}{\sum_{j=1}^{M} (1/Tr(YP^jY^T))^{1/(r-1)}},\ j = \{1, \ldots, M\} \tag{35}$$

5. Experimental results

In this section, we experiment on both synthetic and real-world data sets to evaluate the performance of our method, named FLM. For LEM, LLE, HLLE, LTSA, and our Fusion of local manifolds (FLM) algorithms, we experiment on these data sets to obtain both visualization and quantitative evaluations. We utilize the global smoothness and co-directional consistence (GSCD) criteria [17] to quantitatively compare the embedding qualities of different algorithms: the smaller the value of GSCD, the higher the global smoothness, and the better the co-directional consistence. There are two adjustable parameters in our FLM method, that is, the tuning parameter r and the number of nearest neighbors k. FLM works well when the values of r and k are neither too small nor too large. The reason is that only one local method is chosen when r is too small, while the relative weights of different methods tend to be close to each other when it is too large. As a general recommendation, we suggest to work with $r \in [2, 6]$ and $k \in [0.7\lceil log(N) \rceil,\ 2\lceil log(N) \rceil]$.

5.1. Synthetic data sets

We first apply our FLM to the synthetic data sets that have been commonly used by other researchers: S-Curve, Swiss Hole, Punctured Sphere, and Toroidal Helix. The character of these data sets can be summarized as: general, non-convex, nonuniform, and noise, respectively. In each data set, we have total 1000 sample points, and the number of nearest neighbors is fixed to $k = 10$ for all the algorithms. For the S-Curve and Swiss Hole, we empirically set $r = 2$, and for the Punctured Sphere and Toroidal Helix data sets, we set r=3. **Figures 2–5** show the embedding results of the above algorithms on the four synthetic data sets. Each manifold learning algorithm and the corresponding GSCD result are shown in the title of each subplot. We can evaluate the performances of these methods by comparing the coloring of the data points, the smoothness, and the shape of the projection coordinates with their original manifolds. **Figures 2–5** reveal the following interesting observations.

1. On some particular data sets, the traditional local manifold learning methods perform well. For example, LEM works well on the Toroidal Helix; LLE works well on the Punctured Sphere; HLLE works well on the S-Curve and Swiss Hole; and LTSA performs well on the S-Curve, Swiss Hole, and Punctured Sphere.
2. In general, our FLM performs the best on all the four data sets.

The above consequence is because only partial geometric information of the underlying manifold is learned by each traditional local manifold learning method, while the complementary geometric information learned from different manifold learning algorithms is respected by our FLM method.

5.2. Real-world data set

We next conduct experiments on the isometric feature mapping face (ISOFACE) data set [1], which contains 698 images of a 3-D human head. The ISOFACE data set is collected under different poses and lighting directions. The resolution of each image is 64×64. The intrinsic degrees of freedom are the horizontal rotation, vertical rotation, and lighting direction. The 2-D embedding results of different algorithms and the corresponding GSCD results are shown in **Figure 6**. In the embedding, we randomly mark about 8% points with red circles and attach their corresponding training images. In the experiment, we fix the number of nearest neighbors to $k = 12$ for all the algorithms. We empirically set r in FLM as 4. **Figure 6** reveals the following interesting observations.

1. As we can observe from **Figure 6b** and **c**, the embedding results of LEM and LLE show that the orientations of the faces change smoothly from left to right along the horizontal direction, and the orientations of the faces change from down to up along the vertical direction. However, as we can see at the right-hand side of **Figure 6b** and **c**, the embedding results of both LEM and LLE come out to be severely compressed, and it is not obvious to survey the changes along the vertical direction.
2. As we can observe from **Figure 6d** and **e**, the horizontal rotation and variations in the brightness of the faces can be well revealed by the embedding result of HLLE and LTSA.

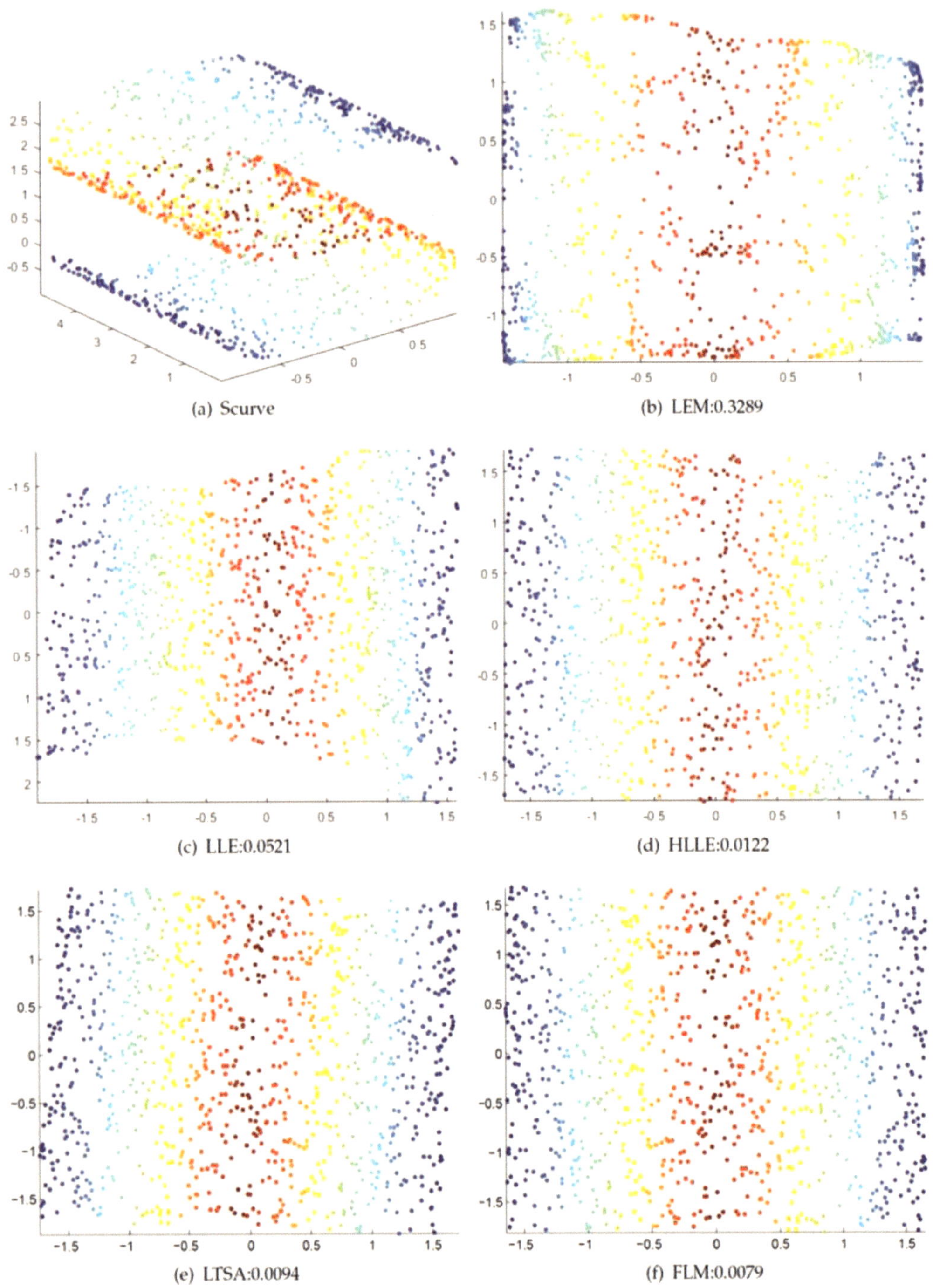

Figure 2. Embeddings of the synthetic manifold S-curve. The title of each subplot indicates the abbreviation of the manifold learning algorithm and the GSCD result. (a) Sample data. The title of subplots (b)-(f) indicates the abbreviation of the the manifold learning algorithm and the GSCD result.

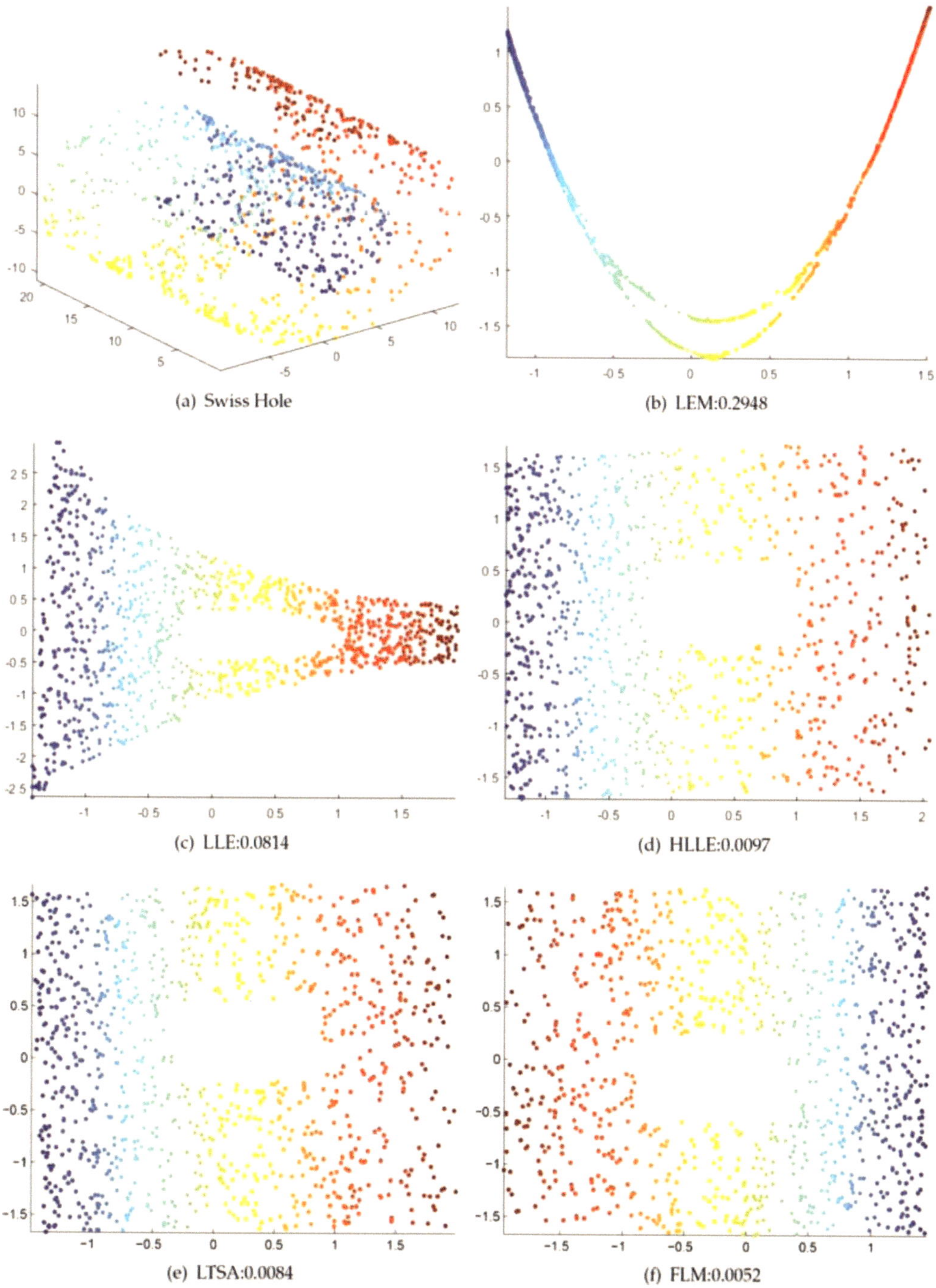

Figure 3. Embeddings of the synthetic manifolds Swiss Hole. The title of each subplot indicates the abbreviation of the manifold learning algorithm and the GSCD result. (a) Sample data. The title of subplots (b)-(f) indicates the abbreviation of the the manifold learning algorithm and the GSCD result.

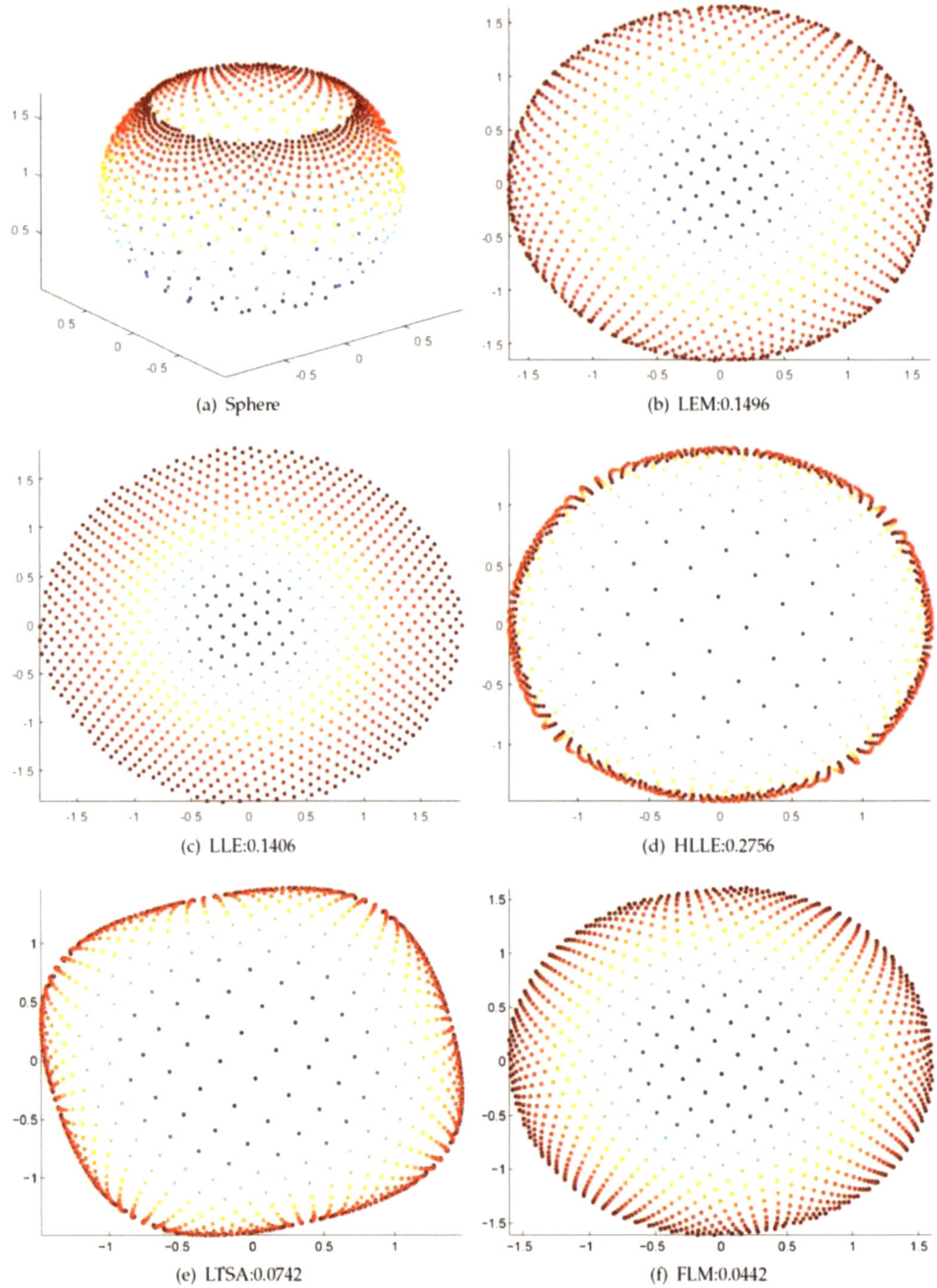

Figure 4. Embeddings of the synthetic manifolds Punctured Sphere. The title of each subplot indicates the abbreviation of the manifold learning algorithm and the GSCD result. (a) Sample data. The title of subplots (b)-(f) indicates the abbreviation of the the manifold learning algorithm and the GSCD result.

http://dx.doi.org/10.5772/66303

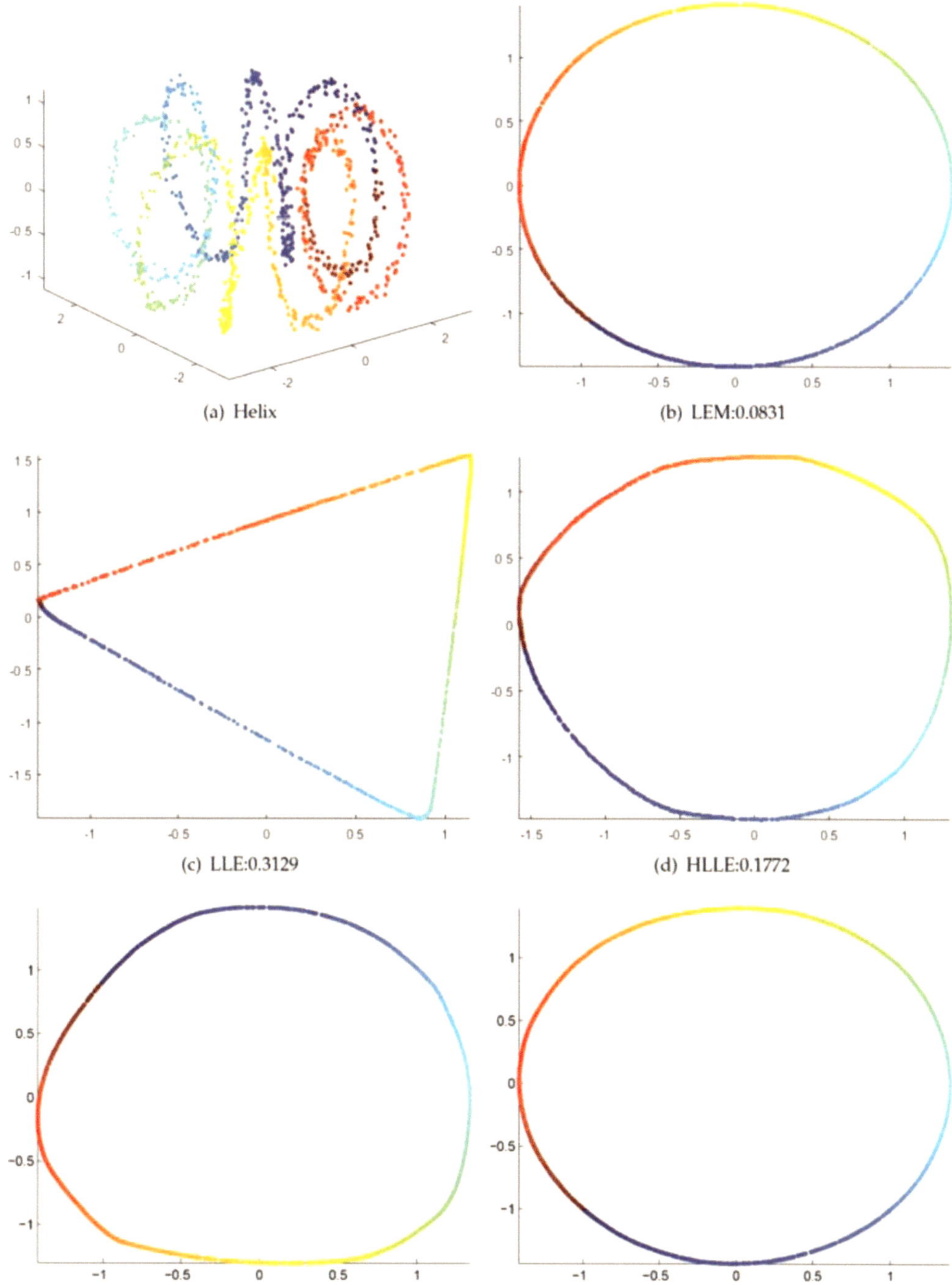

(a) Helix (b) LEM:0.0831

(c) LLE:0.3129 (d) HLLE:0.1772

(e) LTSA:0.1289 (f) FLM:0.0988

Figure 5. Embeddings of the synthetic manifolds Toroidal Helix. The title of each subplot indicates the abbreviation of the manifold learning algorithm and the GSCD result. (a) Sample data. The title of subplots (b)-(f) indicates the abbreviation of the the manifold learning algorithm and the GSCD result.

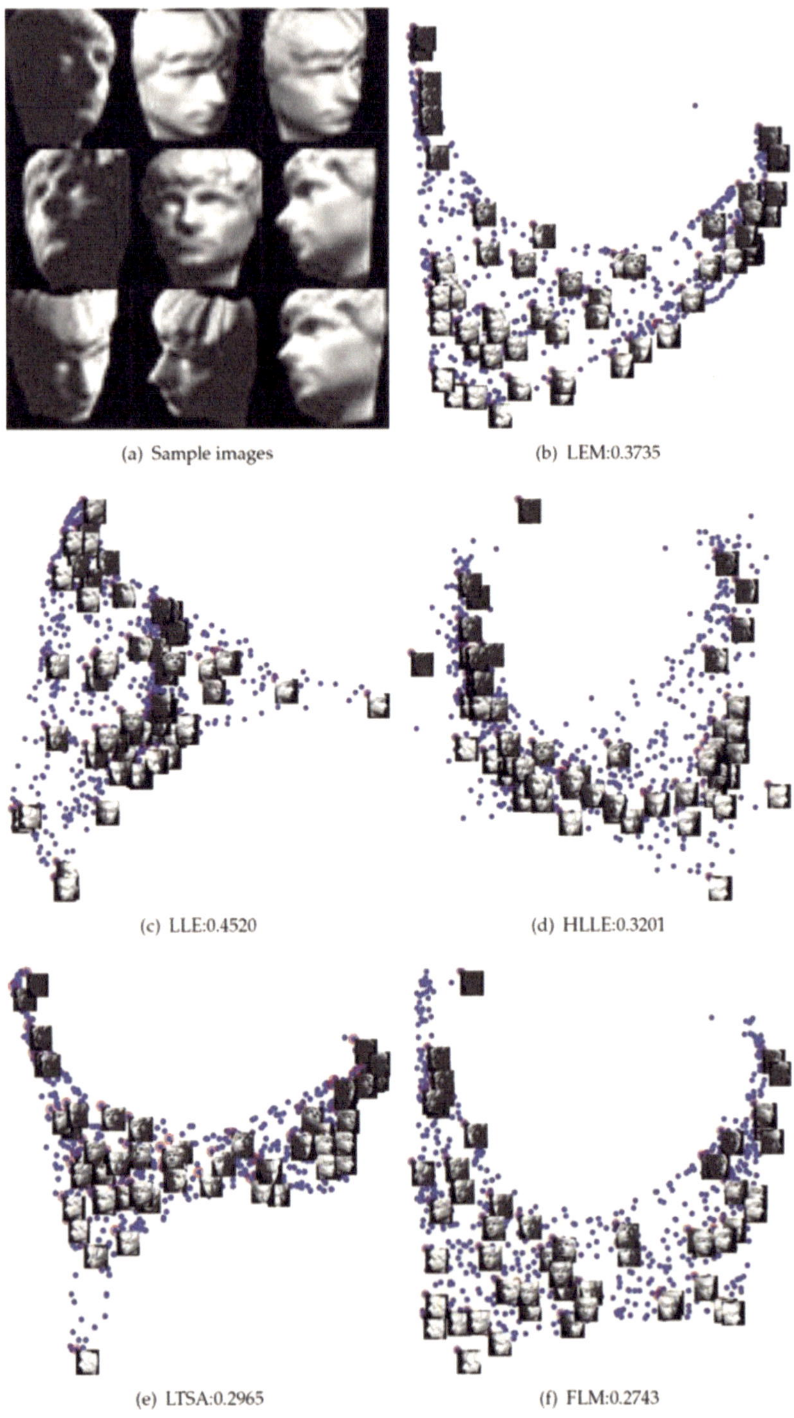

Figure 6. Embeddings of the ISOFACE data set. Subfigure (a) shows nine sample images, and subfigure (b) to subfigure (f) are the embedding results of different manifold learning algorithms. The title of each subplot indicates the abbreviation of the manifold learning algorithm and the GSCD result.

3. As we can observe from **Figure 6f**, orientations of the faces change smoothly from left to right along the horizontal direction, while the orientations of the faces change from down to up, and the light of the faces varies from bright to dark simultaneously along the vertical direction. These results illustrate that our FLM method successfully discovers the underlying manifold structure of the data set.

Our FLM performs the best on the ISOFACE data set, since our method makes full use of the complementary geometric information learned from different manifold learning methods. The corresponding GSCD results further verify the above visualization results in a quantitative way.

6. Conclusions

In this chapter, we introduce an interesting method, named FLM, which assumes a systematic framework to estimate the local objects and align them to reveal a single global low-dimensional coordinate space. Within the framework, we can fuse together the geometric information learned from different local methods easily and effectively to better discover the underlying manifold structure. Experimental results on both the synthetic and real-world data sets show that the proposed method leads to satisfactory results.

Acknowledgements

This work was supported by the Fundamental Research Funds for the Central Universities of China, Natural Science Fund of Heilongjiang Province of China, and Natural Science Foundation of China under Grant No. HEUCF160415, F2015033, and 61573114.

Author details

Xianglei Xing*, Kejun Wang and Weixing Feng

*Address all correspondence to: xingxl@hrbeu.edu.cn

College of Automation, Harbin Engineering University, China

References

[1] Tenenbaum JB, De Silva V, Langford JC. A global geometric framework for nonlinear dimensionality reduction. Science. 2000;290(5500):2319–2323.

[2] Weinberger KQ, Saul LK. Unsupervised learning of image manifolds by semidefinite programming. International Journal of Computer Vision. 2006;70(1):77–90.

[3] Chen J, Ma Z, Liu Y. Local coordinates alignment with global preservation for dimensionality reduction. IEEE Transactions on Neural Networks and Learning Systems. 2013;24(1):106–117.

[4] Belkin M, Niyogi P. Laplacian eigenmaps for dimensionality reduction and data representation. Neural Computation. 2003;15(6):1373–1396.

[5] Roweis ST, Saul LK. Nonlinear dimensionality reduction by locally linear embedding. Science. 2000;290(5500):2323–2326.

[6] Donoho DL, Grimes C. Hessian eigenmaps: locally linear embedding techniques for high-dimensional data. Proceedings of the National Academy of Sciences. 2003;100(10):5591–5596.

[7] Zhang ZY, Zha HY. Principal manifolds and nonlinear dimensionality reduction via tangent space alignment. Journal of Shanghai University (English Edition). 2004;8 (4):406–424.

[8] Hou C, Wang J, Wu Y, Yi D. Local linear transformation embedding. Neurocomputing. 2009;72(10):2368–2378.

[9] Hou C, Zhang C, Wu Y, Jiao Y. Stable local dimensionality reduction approaches. Pattern Recognition. 2009;42(9):2054–2066.

[10] Wang R, Shan S, Chen X, Chen J, Gao W. Maximal linear embedding for dimensionality reduction. IEEE Transactions on Pattern Analysis and Machine Intelligence. 2011;33 (9):1776–1792.

[11] Yan S, Xu D, Zhang B, Zhang HJ, Yang Q, Lin S. Graph embedding and extensions: a general framework for dimensionality reduction. IEEE Transactions on Pattern Analysis and Machine Intelligence. 2007;29(1):40–51.

[12] Jolliffe I. Principal component analysis. Wiley Online Library; Springer-Verlag New York 2005.

[13] Turk M, Pentland A. Eigenfaces for recognition. Journal of Cognitive Neuroscience. 1991;3(1):71–86.

[14] Belkin M, Niyogi P. Towards a theoretical foundation for Laplacian-based manifold methods. Journal of Computer and System Sciences. 2008;74(8):1289–1308.

[15] Chern SS, Chen WH, Lam KS. Lectures on differential geometry. Vol. 1. World Scientific Pub Co Inc; Singapore 1999.

[16] James C. Bezdek, Richard J. Hathaway. Some notes on alternating optimization. In: Advances in Soft Computing. Springer; Springer Berlin Heidelberg 2002. p. 288–300.

[17] Zhang J, Wang Q, He L, Zhou ZH. Quantitative analysis of nonlinear embedding. IEEE Transactions on Neural Networks. 2011;22(12):1987–1998.